T0218335

This series aims to report new developments in mathematical research and teaching – quickly, informally and at a high level. The type of material considered for publication includes:

1. Preliminary drafts of original papers and monographs

2. Lectures on a new field, or presenting a new angle on a classical field

3. Seminar work-outs

4. Reports of meetings

Texts which are out of print but still in demand may also be considered if they fall within these categories.

The timeliness of a manuscript is more important than its form, which may be unfinished or tentative. Thus, in some instances, proofs may be merely outlined and results presented which have been or will later be published elsewhere.

Publication of *Lecture Notes* is intended as a service to the international mathematical community, in that a commercial publisher, Springer-Verlag, can offer a wider distribution to documents which would otherwise have a restricted readership. Once published and copyrighted, they can be documented in the scientific literature.

Manuscripts
Manuscripts are reproduced by a photographic process; they must therefore be typed with extreme care. Symbols not on the typewriter should be inserted by hand in indelible black ink. Corrections to the typescript should be made by sticking the amended text over the old one, or by obliterating errors with white correcting fluid. Should the text, or any part of it, have to be retyped, the author will be reimbursed upon publication of the volume. Authors receive 75 free copies.

The typescript is reduced slightly in size during reproduction; best results will not be obtained unless the text on any one page is kept within the overall limit of 18 x 26.5 cm (7 x 10 ½ inches). The publishers will be pleased to supply on request special stationery with the typing area outlined.

Manuscripts in English, German or French should be sent to Prof. Dr. A. Dold, Mathematisches Institut der Universität Heidelberg, Tiergartenstraße or Prof. Dr. B. Eckmann, Eidgenössische Technische Hochschule, Zürich.

Die „*Lecture Notes*" sollen rasch und informell, aber auf hohem Niveau, über neue Entwicklungen der mathematischen Forschung und Lehre berichten. Zur Veröffentlichung kommen:

1. Vorläufige Fassungen von Originalarbeiten und Monographien.

2. Spezielle Vorlesungen über ein neues Gebiet oder ein klassisches Gebiet in neuer Betrachtungsweise.

3. Seminarausarbeitungen.

4. Vorträge von Tagungen.

Ferner kommen auch ältere vergriffene spezielle Vorlesungen, Seminare und Berichte in Frage, wenn nach ihnen eine anhaltende Nachfrage besteht.

Die Beiträge dürfen im Interesse einer größeren Aktualität durchaus den Charakter des Unfertigen und Vorläufigen haben. Sie brauchen Beweise unter Umständen nur zu skizzieren und dürfen auch Ergebnisse enthalten, die in ähnlicher Form schon erschienen sind oder später erscheinen sollen.

Die Herausgabe der „*Lecture Notes*" Serie durch den Springer-Verlag stellt eine Dienstleistung an die mathematischen Institute dar, indem der Springer-Verlag für ausreichende Lagerhaltung sorgt und einen großen internationalen Kreis von Interessenten erfassen kann. Durch Anzeigen in Fachzeitschriften, Aufnahme in Kataloge und durch Anmeldung zum Copyright sowie durch die Versendung von Besprechungsexemplaren wird eine lückenlose Dokumentation in den wissenschaftlichen Bibliotheken ermöglicht.

Lecture Notes in Mathematics

A collection of informal reports and seminars
Edited by A. Dold, Heidelberg and B. Eckmann, Zürich

110

The Many Facets of Graph Theory

Proceedings of the Conference held at Western Michigan
University, Kalamazoo / MI., October 31 – November 2, 1968

Edited by G. Chartrand and S. F. Kapoor
Western Michigan University, Kalamazoo / MI

Springer-Verlag
Berlin · Heidelberg · New York 1969

© by Springer-Verlag Berlin · Heidelberg 1969. Library of Congress Catalog Card Number 70–101693. Title No. 3266.

These proceedings
are dedicated to
the memory of

PROFESSOR OYSTEIN ORE,

outstanding mathematician and eminent graph theorist, who died unexpectedly in the summer of 1968. Professor Ore, who had retired from Yale University in the spring of 1968, was to have been the principal speaker at this Conference on Graph Theory, and his death deprived the mathematical world of a highly respected colleague.

FOREWORD

This volume constitutes the proceedings of the Conference on Graph
Theory held at Western Michigan University in Kalamazoo, 31 October -
2 November 1968. Its participants represent a diverse spectrum of the
mathematical scale. Those in attendance ranged from graduate students
to outstanding research mathematicians[1], and their contributions to
this volume[2] include a wide variety of expositions, applications of
graph theory, and topics of current research. It is our hope that
this collection of papers will begin to acquaint the reader with some
of "the many facets of graph theory."

As with all such ventures of this nature, the editors are deeply
indebted to many people for the success of this undertaking - not only
for their assistance in planning and conducting the Conference, but
also for their help in the preparation of this volume. We apologize
in advance for any errors or oversights which may exist in the final
edition.

We thank the Department of Mathematics of Western Michigan University for the use of facilities and financial support and Professor
A. Bruce Clarke, Chairman, for his encouragement and counsel before,
during, and after the Conference.

We extend our appreciation to Professor Yousef Alavi for his handling of the arrangements for the Conference and for looking after
many of the details which allowed it to function smoothly.

We are particularly indebted to Mrs. Judith Warriner, who typed
most of the manuscript, for her excellent work and cheerful patience
in seeing it through several revisions.

Credit is also due to Mr. Kenneth Musselman, undergraduate assistant, who helped to conduct the Conference and later to proofread the
manuscripts, and to Mrs. Janet Bunce, who assisted with the typing.

A special debt of gratitude is owed to Mr. Scott Marovich, undergraduate assistant, who not only helped to conduct the Conference and
later to proofread the manuscripts, but who also perpared all drawings
which occur in the various papers. Mr. Marovichwas also instrumental
in the organization of this volume.

Finally, we wish to thank Springer-Verlag for publishing these
proceedings.

July, 1969   G.C.
                                                    S.F.K.

---

[1] We thank R. K. Guy whose paper inspired the title of this volume.

[2] P. Erdös and D. P. Geller were unable to attend the Conference, but
their contributions are included as part of the published record.

TABLE OF CONTENTS

VI

FUNDAMENTAL DEFINITIONS

Introduction. This article introduces the notation and many of the definitions used in the lectures to follow. Ordinarily, each article includes additional definitions pertinent to the particular exposition. The reader who is familiar with the basic terminology of graph theory may, of course, proceed directly to those articles which are of interest to him.

Basic Definitions. Unless otherwise stated, a _graph_ G is a finite, nonempty set V(G) of _vertices_ or _points_ together with a set E(G) consisting of 2-element subsets of V(G); an element {u, v} is called an _edge_ or _line_ and is also denoted by (u, v), (v, u), uv, or vu. The edge uv is said to _join_ the vertices u and v. If e = uv is an edge of a graph G, then u and v are _adjacent vertices_ and e and u (e and v) are _incident_. If e = uv and f = vw are distinct, then e and f are _adjacent edges_. Two graphs G_1 and G_2 are _isomorphic_ (written $G_1 \cong G_2$) if there exists a one-to-one mapping from $V(G_1)$ onto $V(G_2)$ such that two vertices of G_1 are adjacent if and only if the corresponding vertices of G_2 are adjacent.

The number of elements in the vertex set V(G) of a graph G is called the _order_ of G; if G has order one, then G is _trivial_. If v is a vertex of G, then the number of edges of G incident with v is the _degree_ or _valency_ of v and is often denoted deg v or d(v).

A sequence $v_1, v_2, \ldots, v_n$ of distinct vertices of a graph G in which every two consecutive vertices are adjacent is called a v_1-v_n _path_ of G. If there exists a u-v path for every two distinct vertices u, v of a graph G, then G is _connected_. Two u-v paths of G are called _disjoint_ if the paths have no vertices in common, other than u or v. If, in the above sequence, n ≥ 3 and $v_1 v_n$ is also an edge of G, then the sequence $v_1, v_2, \ldots, v_n, v_1$ is a _cycle_ or _circuit_ of G.

Let G be a graph; if V' and E' are subsets of V(G) and E(G), respectively, which together form a graph H, then H is a _subgraph_ of G. If F is a subgraph of G such that S = V(F) and each edge of G which joins two vertices of S is also an edge of F, then F is called the _subgraph_ of G _induced by_ S, and we write F = <S>. A subset S of V(G) such that <V(G) - S> is not connected is called a _cutset_ of G (an _n-cutset_ if |S| = n).

Special Graphs. The _complete graph_ K_p is the graph of order p, every two vertices of which are adjacent. A graph G is _bipartite_ if V(G) can be partitioned into nonempty subsets V_1 and V_2 such that each edge of G joins a vertex of V_1 with a vertex of V_2; G is the _complete bipartite graph_ K(m, n) if G is bipartite, $|V_1| = m$,

$|V_2| = n$, and each vertex of V_1 is adjacent to every vertex of V_2.

A cycle of a graph G is called <u>hamiltonian</u> if it contains every vertex of G; a <u>hamiltonian graph</u> is one which contains a hamiltonian cycle. A connected graph with no cycles is referred to as a <u>tree</u>.

There are numerous ways of associating a graph with a given graph. We describe a few of these which appear in the following lectures.

The <u>complement</u> $\overline{G}$ of a graph G (also denoted C(G)) is that graph such that $V(\overline{G}) = V(G)$ and such that two vertices of $\overline{G}$ are adjacent if and only if they are not adjacent in G.

The <u>line graph</u> L(G) of a graph G, with $E(G) \neq \phi$, is a graph whose vertex set can be placed in a one-to-one correspondence with E(G) such that two vertices of L(G) are adjacent if and only if the corresponding edges of G are adjacent. Similarly, the <u>total graph</u> T(G) of a graph G is defined as that graph whose vertex set can be put in one-to-one correspondence with $V(G) \cup E(G)$ such that two vertices of T(G) are adjacent if and only if the corresponding elements of G are adjacent or incident.

Let G_1 and G_2 be two graphs having disjoint vertex sets. Then the <u>join</u> of G_1 and G_2, often denoted $G_1 + G_2$, is that graph for which $V(G_1 + G_2) = V(G_1) \cup V(G_2)$ and $E(G_1 + G_2) = E(G_1) \cup E(G_2) \cup$ E', where E' consists of all edges of the type $v_1 v_2$, $v_1 \varepsilon V(G_1)$ and $v_2 \varepsilon V(G_2)$.

<u>Directed Graphs</u>. A <u>directed graph</u> (or digraph) D is a finite non-empty set V(D), whose elements are called <u>vertices</u> or <u>points</u>, together with a set E(D) of ordered pairs of distinct elements of V(D); the elements of E(D) are often called <u>directed edges</u>, <u>directed lines</u>, or <u>arcs</u>. If e = (u, v) is an arc of a digraph D, then u is <u>adjacent to</u> v and v is <u>adjacent from</u> u; in this case u is called the <u>initial vertex</u> or <u>tail</u> of the arc e and v is the <u>terminal vertex</u> or <u>head</u>.

The number of vertices which are adjacent to a vertex v in a digragh D is called the <u>indegree</u> of v (sometimes denoted id v) while the number of vertices which are adjacent from v is called the <u>outdegree</u> of v (denoted od v). On occasion, the <u>degree</u> of a vertex v in a digraph D is defined as the sum of its indegree and outdegree.

A <u>tournament</u> is a digraph D with the property that for every two distinct vertices u and v, exactly one of (u, v) and (v, u) is an arc of D.

G.C.

S.F.K.

GRAPHS AND BINARY RELATIONS[1]

Martin Aigner, University of North Carolina

I. Introduction. Much attention has recently been paid to character-izations of important classes of graphs by means of certain subgraphs which they must not contain. As the first characterization of this type was given by Kuratowski for the class of planar graphs, we will refer to them as Kuratowski characterizations. Besides planar graphs and related structures (see [3]) other interesting classes of graphs G(R) which permit such characterizations arise by considering cer-tain binary relations R on a set A. For example, the graphs of strict (irreflexive) partial orders (PO-graphs) have been character-ized by Gilmore-Hoffman [7], Ghouila-Houri [5, 6] and Gallai [4]; the graphs of semi-orders (SO-graphs) and indifference systems (I-graphs) have been characterized by Roberts [13]; and the graphs of interval-systems (IV-graphs) have been characterized by Lekkerkerker-Boland [10], Gilmore-Hoffman [7] and Roberts [13]. All these characteriza-tions are of the Kuratowski type and will be summarized in the next section. In fact, all these types of graphs are special cases of the class of perfect graphs which, too, possess Kuratowski characteriza-tions [14].

In this paper, we discuss three sets of problems for the four aforementioned classes of graphs.

A. Extremal problems: Determination of all minimal Non PO-graphs, Non SO-graphs, etc., such that the deletion of any line, or any point, gives rise to a PO-graph, SO-graph, etc. We will call these the line problem, and point problem, respectively.

B. Problems concerning incidence patterns P (see [8]): Character-izations of graphs G such that P(G) is a PO-graph, etc. In this note, we will confine ourselves to a discussion of the line-graph L(G).

C. Uniqueness: When does G(R) completely determine R, i.e., when does G(R) = G(R') imply R = R' or R' = R^I, the inverse relation of R? (R, R' are relations of the same type, of course.)

References to the corresponding problems for planar graphs and related matters can be found in [3, 12, 16, 17]. As to the uniqueness problem, we may further ask the question: Under what conditions does a planar graph permit essentially only one embedding into the plane? 3-connectedness is known to be a sufficient condition, and it will be interesting to compare it with the sufficient condition for PO-graphs to be uniquely partially orderable which we will derive in Section 5.

[1]Research partially sponsored by the Air Force Office of Scientific Research and Office of Aerospace Research, U.S. Air Force, under AFOSR Grant No. 68-1406.

II. <u>Preliminaries</u>. Except for the case of SO-graphs, we shall not permit graphs with isolated points. Given a graph G, we write a ~ b if and only if (a,b) ∈ E(G) for a, b ∈ V(G).

In Figure 1, three graphs are displayed which will appear throughout the paper. We shall refer to them as G_1, G_2, G_3.

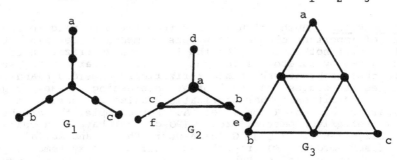

Figure 1

Notice that $G_2 = L(G_1)$, $G_3 = L(G_2)$, and $G_3 = \overline{G}_2$, three facts which will prove important later on.

<u>Definition 1</u>. A <u>generalized</u> <u>path</u> of length k-1 in a graph G is a progression of vertices a_1, a_2, ..., a_k, $a_i ∈ V(G)$, $(a_i, a_{i+1}) ∈$ E(G), such that no two ordered pairs (a_i, a_{i+1}), (a_j, a_{j+1}) are the same (i.e., an edge may be traversed twice, but at most once in either direction). We speak of a <u>generalized</u> <u>cycle</u>, if the vertex progression is a generalized path with $a_1 = a_k$.

<u>Definition 2</u>. Given the generalized path a_1, ..., a_k, an edge (a_i, a_j) with $|i - j| ≥ 2$ is called a <u>chord</u>; if $|i - j| = 2$ we speak of a <u>triangular</u> <u>chord</u>. Similar definitions hold for generalized cycles, modulo k - 1.

Given the binary relation R on a finite set A, we define the associated graph G(R) by V(G(R)) = A, and (x,y) ∈ E(G(R)) if and only if (xRy) ∨ (yRx) for x,y ∈ A, and the resulting structure is denoted by (A,R). If there is no danger of ambiguity, we may sometimes use the same letter G for G = (A,R) and its associated graph G = G(R). Also, G(R) is undirected, but possesses loops if and only if R is reflexive. For obvious reasons, therefore, we shall henceforth restrict our attention to irreflexive relations.

1. <u>Partial orders</u>. As we agreed to rule out loops, the characterization question in this case reads as follows: Under what conditions on G is it possible to direct the edges in such a way that the resulting orientation is transitive. The answer is provided by the following:

<u>Theorem 1 [4, 5, 6, 7]</u>. G is a PO-graph if and only if it contains no generalized cycle of odd length without triangular chords.

It is easily seen, (using the fact that two edges with a common endpoint, but which are not contained in a triangle, must receive opposite directions), that G_2 and G_3 are both non PO-graphs whereas G_1 is a PO-graph. As an example, a generalized cycle as required in the theorem is: G_2 is given by a, b, e, b, c, f, c, a, d, a.

2. Indifference systems.

Definition 3. Given the binary relation I on A, (A,I) is called an indifference system if there exists a real-valued function f on A and a real number $\delta > 0$ such that x I y if and only if $|f(x) - f(y)| \le \delta$. We call f a defining function for (A,I).

The characterization problem then consists of determining under what conditions on G we can find a defining function f on V(G).

Theorem 2 [13]. G is an I-graph if and only if it does not contain $K(1,3)$, G_2, G_3, C_k $(k \ge 4)$ as full subgraphs.

3. Semi-orders.

Definition 4 [11]. The binary relation R is called a semi-order on A if and only if for all x, y, z, w $\in$ A,
   a) x R x,
   b) (x R y $\wedge$ z R w) $\Rightarrow$ (x R w $\vee$ z R y),
   c) (x R y $\wedge$ y R z) $\Rightarrow$ (x R w $\vee$ w R z).

A semi-order is clearly irreflexive, asymmetric and transitive, hence a (strict) partial order. Equivalently, any SO-graph is a PO-graph. Using results from [15] and [13], which show that a graph G is a SO-graph if and only if its complement C(G) is an I-graph, we obtain the following:

Theorem 3 [13, 15]. G is a SO-graph iff it does not contain $K_3 +$ K_1, G_2, G_3, $K_2 + K_2$, C_5 or the complement of C_k $(k \ge 6$, without chords) as full subgraphs.

4. Interval systems.

Definition 5. Given a binary relation R on A, (A,R) is called an interval system if there exists an assignment T of non-empty intervals of the real line to the elements of A such that x R y if and only if $T(x) \cap T(y) \ne \phi$ for x, y $\in$ A.

For IV-graphs, two different characterizations of the Kuratowski type exist. Before we state them we need the following:

Definition 6. Three points x, y, z in a graph G are said to form an asteroidal triple (A-triple) $\langle x, y, z \rangle$ if there exist paths P_{xy}, P_{yz}, P_{zx} from x to y, y to z, and z to x, respectively, such that x is not adjacent to any point of P_{yz}, y is not adjacent to any point of P_{zx}, and z is not adjacent to any point of P_{xy}.

Theorem 4 [7]. G is an IV-graph iff it contains no C_4 as full sub-graph, and $C(G)$ is a PO-graph.

In [13], it is shown that I-graphs are exactly those IV-graphs for which an assignment T exists such that $T(x) \not\subset T(y)$ for all x, y. From this result, it is then easy to deduce that a graph G is an I-graph if and only if it is an IV-graph without $K(1,3)$'s as full subgraphs. Hence we have the following:

Corollary 1 [13]. G is an I-graph if and only if it contains no C_4 or $K(1,3)$ as full subgraphs, and its complement $C(G)$ is a PO-graph.

Theorem 5 [10]. G is an IV-graph if and only if it contains no C_k (k ≥ 4) as a full subgraph, and also no asteroidal triple.

We note that all three graphs of Figure 1 contain A-triples, namely ⟨a, b, c⟩, ⟨d, e, f⟩, ⟨a, b, c⟩, respectively; hence none of these graphs is an IV-graph. Furthermore, a cycle C_k without chords plainly contains an A-triple for k ≥ 6, so we can put Theorem 5 in a more compact form: G is IV-graph if and only if it contains no C_4 or C_5 as full subgraphs, and no A-triple.

III. <u>Extremal Configuration</u>. It is immediately clear that for any binary relation R, a graph G is a R-graph if and only if every full subgraph of G is an R-graph; or, equivalently, if a full sub-graph of G is a non R-graph, then so is G. Hence it makes sense to ask what are the smallest non R-graphs in the sense that the de-letion of any line, or any point, results in an R-graph. We will re-fer to these graphs as minimal non R-graphs.

A. <u>The Line Problem</u>.

1. <u>Partial orders</u>. According to Theorem 1, a non PO-graph must contain a generalized odd cycle without triangular chords which we will call a GH-cycle for the remainder of the paper. We denote by {a, b, ..., c, d} the generalized cycle $a = a_1$, $b = a_2$, ..., c = a_{k-2}, d = a_{k-1}, with $a_1 = a_k = a$. We speak of the sequence of ver-tices a, b, ... to indicate the direction in which we run through the cycle. (For example, the sequence a, b is different from the edge (a,b).) Subcycles of a given generalized cycle are denoted in the same fashion provided there is no danger of ambiguity. (For ex-ample, the subcycle {b, ...} in the example above means the given cycle minus the first vertex a and the last two vertices c, d (b, of course, must be adjacent to the vertex immediately preceding c).)

Lemma 1. An arbitrary non PO-graph G must contain a block which to-gether with its outgoing edges does not admit a partial order.

Proof. The algorithm designed by Gilmore and Hoffman allows us to start with any particular edge (or for that matter, with any PO-sub-graph of G) in order to construct a partial ordering of the points. Hence, if all the blocks plus their outgoing edges are PO-graphs, we may start with anyone of them and then keep on orienting the edges. Since by definition of a block, we never return to the same block once we leave it, the algorithm clearly yields a PO-graph.

We remark that we cannot dispose of the condition "with its outgoing edges", as is illustrated by the graph G_2. A trivial corollary of Lemma 1 is the fact that all forests are PO-graphs. In view
of Lemma 1, henceforth we will confine ourselves to blocks plus possible outgoing edges. In Lemmas 2-5, we will study a shortest GH-
cycle C of a non PO-graph G. We run through C in one of the two
possible directions, but keep the direction fixed once we have chosen
it. If the vertex b follows the vertex a, we call a the predecessor of b, b the successor of a, and indicate this fact by
a, b.

Lemma 2. Suppose the vertex a appears more than once in $C = \{a,$
$b, \ldots, c, a, d, \ldots, e\}$. Suppose without loss of generality that
$C' = \{a, b, \ldots, c\}$ is of odd length, then
   (i) $b \sim c$,
  (ii) $b \sim d$, $c \sim e$, unless $C = \{a, b, \ldots, c, a, d = e\}$,
 (iii) $d \sim e$, unless $C'' = \{a, d, \ldots, e\}$ has length 4.

Proof. C' is a generalized cycle of odd length, which together with
the hypothesis on C implies (i). If $b \not\sim d$ or $c \not\sim e$ and C'' is
of length at least 4, then the generalized cycle $\{a, b, \ldots, c, a,$
$d\}$ or $\{a, b, \ldots, c, a, e\}$ (after deletion of possibly duplicated
edges), respectively, would be shorter than C, which is a contradiction. To prove (iii), we assume C'' has length greater than 4,
then by considering the generalized cycle $\{a, b, \ldots, c, a, d, a, e\}$
(again after deletion of duplicated edges) we infer $d \sim e$, making
use of the hypothesis on C.

Lemma 3. If for two vertices a and b, C contains both sequences
a, b and b, a, then they must be consecutive sequences, i.e., C
contains a, b, a or b, a, b.

Proof. Let $C = \{a, b, \ldots, b, a, \ldots\}$. Then we may assume, without
loss of generality, that $C' = \{b, \ldots\}$ and hence $C'' = \{a, b, \ldots,$
b$\}$ are of odd length. But this would clearly contradict Lemma 2 (i).

Lemma 4. Let G be a minimal non PO-graph and C as before, then
every edge of G must appear in C at least once.

Proof. The deletion of any edge not in C would not alter the character of G as to partial orderings, in contradiction to the minimality of G.

Lemma 5. Let G be a minimal non PO-graph and C a shortest GH-
cycle. Suppose C contains the sequence a, b, c, d, with $a \sim d$,
then $a = c$ or $b = d$.

Proof. We proceed to prove the assertion by contradiction. Since
$a \sim d$, C contains the edge (a,d). Assume first that $C = \{a, b,$
$c, d, \ldots, a, d, \ldots\}$. We have to consider two cases depending on
whether $C' = \{d, \ldots, a\}$ has odd or even length. In the first case,
Lemma 2 (ii) would imply $a \sim c$, contradicting the fact that C
does not contain any triangular chords. In the latter case, we have
$C = \{a, d, \ldots, a, b, c, d \ldots\}$ with $C'' = \{a, d, \ldots\}$ being a generalized cycle of odd length. By appealing to Lemma 2 (ii) again, we
conclude $b \sim d$, a contradiction. The case where C contains d
as predecessor of a can be settled in an analogous fashion, thus
the edge (a,d) must appear within the sequence a, b, c, d, and

the lemma follows.

Every simple odd cycle without chords of length at least 5 **ob**viously is a minimal non PO-graph (deletion of any edge produces a tree), as are the graphs G_2 and G_3. The following theorem makes the converse assertion that these are all the minimal graphs.

Theorem 6. The (line-) minimal non PO-graphs are the simple odd cycles of length at least five without chords and the graphs G_2 **and** G_3.

Proof. Let G be an arbitrary minimal non PO-graph and C a short**est** GH-cycle. If no vertex of G appears more than once in C, then C clearly represents a simple cycle of odd length. Since by Lemma 4, C must contain all the edges of G, there can be no **chords** in C, and we obtain the first class of the above mentioned graphs.

Suppose now there are points that occur at least twice in C, then if x is such a vertex appearing, say, k times, we can **think** of C as the union of k cycles, each starting and terminating at x. Since C is of odd length, at least one of these cycles must also have odd length. Let us denote by $C(x)$ one of these cycles of shortest odd length. In the set of points appearing at least twice, choose the point a such that $C(a) = C'$ is a cycle of shortest length among all $C(x)$, call the complementary cycle C'', and let $C = \{a, b, c, ..., d, e, a, f, h, ..., h, g\}$ with $C' = \{a, b, c, ..., d, e\}$ and $C'' = \{a, f, ..., g\}$. If we can show that G contains either G_2 or G_3 as a full subgraph, then the theorem will follow. By the construction of C', it is clear that whenevery a vertex appears more than once in C', the number of edges between the two occurrences is even, a fact which will be used extensively in the sequel. In C', we have $b \sim e$ (by Lemma 2 (i)), and by the minimality of $C' = C(a)$ we infer $c \neq d$, since otherwise $C(c)$ would be shorter than $C(a)$. Thus a, b, c, d, e are 5 distinct vertices.

Case a. $f = g$. Here Lemma 3 implies $C'' = \{a, f\}$, and, by the minimality of C' again, we have $f \not\sim c$, $f \not\sim d$. Furthermore, we note $f \neq b$, $f \neq e$ and clearly $f \not\sim b$, $f \not\sim e$, and so the following situation results:

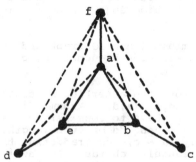

Figure 2

where the broken lines indicate that these edges are missing. To
show that [a, b, c, d, e, f] induces G_2, we have to demonstrate
the absence of the 3 edges (b, d), (c, e), (c, d). The first two
are missing because they cannot be in C' (they would violate the
minimality of C'), and C" only consists of a, f, a. Finally, if
(c, d) were in G, we could delete it and still retain a Non PO-
graph, namely G_2, thus contradicting the hypothesis on G.

Case b. f ≠ g. The edge (b, e) must be contained in C', since
otherwise the deletion of (b, e) would yield the GH-cycle {a, b,
c, ..., d, e}, hence G would not be minimal. Now we may assume,
without loss of generality, that e is the successor of b in C',
since if we have the sequence y, e, b, y' in C', then the generalized cycles {a, b, y', ..., d, e} and {a, b, c, ..., y, e} are
both of odd length (because of the minimality of C' again) and we
either have y = b or y' = e, in which cases e is the successor
of b, or a ~ y, a ~ y' with both edges in C". In the latter
case, application of Lemma 2 readily yields y ~ y', which in turn
implies y = b or y' = e by Lemma 5.

Assume then C' contains the sequence x, b, e, x'. It is our
goal to prove that C' = {a, b, c, b, e, d, e}. To this end, it
suffices to show c = x and d = x', as Lemma 3 clearly indicates.
We assume, therefore, without loss of generality, c ≠ x. (The case
d ≠ x' can be dealt with in an analogous fashion.) We are faced
with four possibilities as to whether x = e or not, and x' = d or
not, a contradiction arising in each of these four cases. A proof is
omitted here, but the details can be found in [1].

The situation at this stage is indicated in Figure 3.

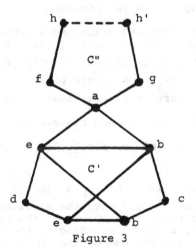

Figure 3

Since f ≠ g, C" is of length at least 4. Using Lemma 2, we
conclude f ~ b, g ~ e and if either one of the edges (f, b), (g,e)
were outside C', we clearly would obtain a GH-cycle after deleting
this edge, thus contradicting the minimality of G. Hence we infer
f = e and g = b (since f ≠ c and g ≠ d). Next we note h ≠ a,
b, e and h' ≠ a, b, e, furthermore h ≠ d and h' ≠ c because of

Lemma 3, and finally $h \neq c$ and $h' \neq d$, since the opposite would
contradict what we just proved about a shortest odd cycle among all
$C(x)$. Lemma 2 applied to the cycle $\{f = e, h, \ldots, c, b, e, d\}$
yields $h \sim d$, and application to the cycle $\{f = e, h, \ldots, c, b\}$
gives $h \sim b$. Similarly we obtain $h' \sim c$ and $h' \sim e$. Now let us
finally consider the edge $(h', e) \in C''$. Using once again Lemma 2,
it is easily seen that either $a \sim h'$ or $h = h'$. As the first possibility cannot occur, we conclude $h = h'$, and the vertices a, b,
c, d, e, h induce G_3.

    2. <u>Indifference systems and semi-orders</u>. According to Theorems
2 and 3, non I-graphs or non SO-graphs must contain certain specified
full subgraphs, hence we merely have to list those forbidden subgraphs
and determine which of them are minimal in the sense explained above.

<u>Theorem 7</u>. The (line-) minimal non I-graphs are $K(1,3)$ and C_k
$(k \geq 4)$ without chords.

<u>Theorem 8</u>. The (line-) minimal non SO-graphs are $K_3 + K_1$ and
$K_2 + K_2$.

    3. <u>Interval systems</u>. First we make the trivial observation
that a non IV-graph must contain at least one component which is a
non IV-graph, and hence we may confine our discussion to connected
graphs.

    Let us introduce at this point the concept of a <u>pendant edge</u> and
of a <u>star graph</u> S_n (sometimes called claw) of order n. We call
(a, b) a pendant edge if at least one of a and b has degree 1.
A star graph S_n consists of a vertex x, called the center, plus
n pendant edges having x as an endpoint.

<u>Lemma 6</u>. Let G be a connected graph and suppose we can find a se-
quence of complete subgraphs $L_1, \ldots, L_t$ (in this lemma, K_2 only
appears among the L_j's if it is not a pendant edge) in G, such
that $V(L_j) \cap V(L_{j+1}) = \{x_j\}$, a single point, and $V(L_j) \cap V(L_{j'}) =$
$\emptyset$ for $j' > j+1$, and further suppose that $E(G)$ consists of
$\bigcup_{j=1}^{t} E(L_j)$ plus possible star graphs centered at $x_1, \ldots, x_{t-1}$ and
some point $x_0 \neq x_1 \in L_1$ and $x_t \neq x_{t-1} \in L_t$. Then G is an IV-
graph.

<u>Proof</u>. Since there are plainly no C_k's $(k \geq 4)$ contained as full
subgraphs in G, we have to disprove the existence of an A-triple in
order to establish the result. Suppose, to the contrary, that there
exists an A-triple $\langle a, b, c \rangle$. According to the hypothesis, any
point x in G is in exactly one of the L_i's, say L_{i_x}, unless
x is one of the connecting points x_j or one of the outer points of
a star graph centered at x_k. If $x = x_j$, assign L_j to x, if x
is an outer point and adjacent to x_k, assign L_k (in case $k = 0$,

we assign L_1). Let us assume, without loss of generality, that $i_a \leq i_b \leq i_c$ for out A-triple $\langle a, b, c \rangle$. By the setup of G, it is clear that b must be adjacent to at least one point on any path connecting a and c, unless b is an outer point in a star graph centered at x_0 or x_t with $a \in L_1$ or $c \in L_t$, respectively. But, if this is the case, then there is no path connecting b with c (or b with a) which does not contain a point adjacent to a (or c).

<u>Corollary 2</u>. A tree is an IV-graph if and only if it does not contain G_1 as a subgraph.

<u>Proof</u>. We only have to verify the sufficiency part, but under the hypothesis, the tree is easily seen to be a graph of the type described in the previous lemma.

<u>Theorem 9</u>. The (line-) minimal non IV-graphs are G_1, G_2 and C_k ($k \geq 4$) without chords.

<u>Proof</u>. First we observe, using Corollary 2, that G_1, G_2 and C_k are minimal non IV-graphs. Let G now be an arbitrary minimal non IV-graph. If G contains a C_k for $k \geq 4$ as full subgraph, then $G = C_k$. In the case where G is a tree, we apply Corollary 2 and obtain $G = G_1$. Suppose finally that G contains no C_k as full subgraph for $k \geq 4$, but does contain triangles. Since G is minimal, it cannot contain a C_4 with just one chord, since we could delete this chord and still retain a non IV-graph. Using this fact, we conclude that if a point x is not in a maximal complete subgraph K_i ($i \geq 3$), it can be adjacent to at most one point of K_i. This in turn implies that if three or more points of a K_i ($i \geq 3$) are adjacent to points outside K_i, then G contains G_2 as a full subgraph; thus $G = G_2$. So let us assume that any one of the complete maximal subgraphs K_i contains at most two points which are adjacent to points not in K_i. But in this case, it is shown with little difficulty (using the fact that two such complete subgraphs intersect in at most one point), that G is either not minimal or has a structure similar to that described in Lemma 6 and is an IV-graph.

B. The Point Problem.

In the case of partial orders, the question was discussed in great depth by Gallai [4]. He calls the (point-) minimal graphs irreducible, and succeeds in determining all of them.

In the case of indifference systems and semi-orders, we again just have to determine which of the forbidden subgraphs are minimal.

<u>Theorem 10</u>. The (point-) minimal non I-graphs are $K(1,3)$, G_2, G_3 and C_k ($k \geq 4$) without chords.

Theorem 11. The (point-) minimal non SO-graphs are $K_3 + K_1$, G_2, G_3 and the complements of C_k ($k \geq 4$) without chords.

Turning to interval systems, we readily verify that C_k without chords for $k \geq 4$ is (point-) minimal, as are G_1 and the graphs displayed in Figure 4. We make the converse assertion and proceed to prove

Theorem 12. The (point-) minimal non IV-graphs are C_k ($k \geq 4$) without chords, G_1, and the graphs of Figure 4.

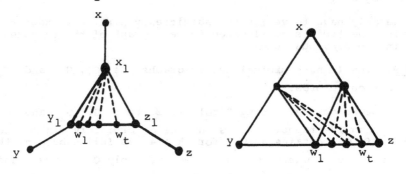

Figure 4

In the first class of graphs in Figure 4, the set of the w_i's may be empty, in which case G_2 results. In the second class t may equal 1, in which case we obtain G_3.

Proof. In view of the remark preceding the theorem, we may confine our discussion to minimal graphs G without C_k's ($k \geq 4$) as full subgraphs. It then follows from Theorem 5 that G must contain an A-triple $\langle x, y, z \rangle$. Let us choose the paths P_{xy}, P_{yz}, P_{zx} to be of shortest possible length, i.e., all their vertices are distinct and there are no chords.

Since G is minimal, it must equal the subgraph induced by the set $V(P_{xy}) \cup V(P_{yz}) \cup V(P_{zx})$. Let us agree on the following nota-tion for this proof: P_{ab} indicates that we run through the path P_{ab} from a to b. If $u, v \in V(P_{xy})$, then by $[u, v]$ we mean the set of all vertices in P_{xy} between u and v, including u, v. Consider the two paths P_{xy}, P_{xz}. They have at least the vertex x in common. Define x_1 as the last vertex which both paths have in common such that $[x, x_1] \subset V(P_{xy}) \cap V(P_{xz})$. Define y_1 and z_1 in the same manner, then we are given the following situation (Figure 5).

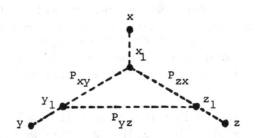

Figure 5

We remark, at this point, that x, y, z are only adjacent to their immediate successors on the paths P_{xy}, P_{xz}; P_{yx}, P_{yz}; P_{zx}, P_{zy}, respectively.

<u>Case a</u>. Suppose there are vertices in $[x, x_1]$ which are adjacent to vertices in $[y_1, z_1] - \{y_1, z_1\}$. In this set, let x^* be the first one. We then wish to prove $x^* = x_1$ and $x \sim x_1$. First we note that $x^* \neq x$. Further, if $[x, x^*] - \{x, x^*\} \neq \emptyset$, we could delete x and the edge incident with it and retain a non IV-graph. Hence x^* is the successor of x, and thus $x \sim x^*$. Suppose now $x^* \neq x_1$, and let $w \in [y_1, z_1] - \{y_1, z_1\}$ so that $x^* \sim w$. We construct two paths P'_{xy} and P'_{xz} as follows. We go from x to x^*, then to w and from there along the path P_{zy} to y or P_{yz} to z, respectively. With the three paths P'_{xy}, P_{yz}, P'_{zx}, $\langle x, y, z \rangle$ clearly forms an A-triple again. Consider now the successor y_2 of y_1 in P_{yx}. Then y_2 is in $[y_1, x_1]$ and it cannot be in $[y_1, z_1]$ (since P_{yz} contains no chords). Hence we could delete y_2 and the edges incident with it without destroying the A-triple $\langle x, y, z \rangle$ constructed above. So we conclude that $x^* = x_1$, and furthermore, x_1, y_1 and z_1 are distinct and $x_1 \sim y_1$, $x_1 \sim z_1$, and thus we obtain the first type of graphs (except G_2) of Figure 4.

Since the assumption that there are vertices in $[y, y_1]$ or $[z, z_1]$ which are adjacent to vertices in $[x_1, z_1] - \{x_1, z_1\}$ or $[x_1, y_1] - \{x_1, y_1\}$, respectively, leads back to Case a we now assume

<u>Case b</u>. There are no vertices in $[x, x_1]$, $[y, y_1]$, $[z, z_1]$ which are adjacent to vertices in $[y_1, z_1] - \{y_1, z_1\}$, $[x_1, z_1] - \{x_1, z_1\}$, $[x_1, y_1] - \{x_1, y_1\}$, respectively. Suppose first $|[x, x_1]| \geq 3$

and let x' be the successor of x in $[x, x_1]$. Then after deletion of x plus the edge (x, x'), $\langle x', y, z \rangle$ form an A-triple unless $x' \sim y_1$ or $x' \sim z_1$, which implies $x_1 = y_1$ or $x_1 = z_1$.
Assume without loss of generality that $x_1 = y_1$. It is now readily
seen that $|[x, x_1]| = 3$, $|[y, x_1]| = 3$, $|[z, x_1]| = 3$ and G_1
results.

Suppose next $|[x, x_1]| = |[y, y_1]| = |[z, z_1]| = 2$. If
$|[x_1, y_1]| = |[x_1, z_1]| = |[y_1, z_1]| = 2$, then we clearly obtain G_Z
Assume that $[x_1, y_1] - \{x_1, y_1\} \neq \phi$, and let u be the successor
of y_1 in P_{yx}. Since by hypothesis there are no C_k's $(k \geq 4)$
without chords in G and since P_{xy}, P_{yz}, P_{zx} contain no chords, u
must be adjacent to the successor v of y_1 in P_{yz}. But this implies $[y_1, z_1] - \{y_1, z_1\} \neq \phi$. Similarly we have $[x_1, z_1] - \{x_1,$
$z_1\} \neq \phi$. We now construct the paths P''_{xy}, P''_{xz} by going from x to
x_1, along P_{xy} to u, to v and from there along P_{zy} to y or
P_{yz} to z, respectively. Given the paths P''_{xy}, P_{yz}, P''_{zx}, $\langle x, y, z \rangle$
again constitutes an A-triple. But now we could delete the successor
of z_1 in P_{zx}, thus violating the minimality of G.

Suppose finally $x = x_1$. Then $|[x, y_1]| \geq 3$, and so by the
above argument we also have $|[x, z_1]| \geq 3$, $|[y_1, z_1]| \geq 3$. This
now implies $y = y_1$ and $z = z_1$, since $\langle x_1, y_1, z_1 \rangle$ is clearly an
A-triple. Denote the points of $P_{yz} - \{y, z\}$ by $w_1, \ldots, w_t$ in
this order, i.e., w_1 is the successor of y, w_t is the predecess-
or of z. Now w_1 is adjacent to the successor u of y in P_{yx}.
Assume $w_1 \neq w_t$, then constructing the path $P_{w_1 x}$ by going from w_1
to u and then along P_{yx} to x, we see that we could delete y
and its two incident edges and still have an A-triple, namely $\langle w_1,$
$x, z \rangle$ with the paths $P_{w_1 x}$, P_{xz}, P_{zw_1}, unless w_1 is adjacent to
some vertex of P_{xz}. If this is the case, then it is easily shown
that there can only be one point in $[x, z] - \{x, z\}$, i.e., $|[x,z]|$
= 3. Similarly we obtain $|[x, y]| = 3$, and the second class of
graphs (except G_3) of Figure 4 results. If lastly $|[x, y]| =$
$|[x, z]| = |[y, z]| = 3$, then we obtain G_3, and the proof is complete.

IV. <u>The Line Graph</u>. For our types of binary relations R we now
ask the question: What is a graph G like whose line graph $L(G)$
is a R-graph? We recall the following characterization of graphs H
which are the line graph of some other graph G [9]: H is the line
graph of some graph G if and only if there exists an edge-disjoint
collection of complete graphs L_i covering H such that each vertex of H belongs to at most two members of the collection. Furth-

ermore, we obtain G such that $H = L(G)$ by defining vertices corresponding to L_i and joining two vertices if and only if their corresponding complete subgraphs in H have a point in common; if there
are k points in L_i which are not in another complete subgraph L_j,
then we attach k pendant edges to the vertex in G corresponding
to L_i.

With this preparation, we immediately have the following:

<u>Lemma 7</u>. Let $H = L(G)$, then H contains, respectively, C_k
($k \geq 4$), G_2, G_3, $K_2 + K_2$, as full subgraphs if and only if G contains, respectively, C_k ($k \geq 4$), G_1, G_2, $K(2,1) + K(2,1)$ as (not
necessarily full) subgraphs. H cannot contain $K(1,3)$ as full subgraph.

<u>Theorem 13</u>. Let G be a graph, then $L(G)$ is a PO-graph if and on-
ly if
    (i) G contains no odd cycle of length ≥ 5,
   (ii) G contains no triangle with edges leading from the three
           vertices into distinct blocks, or with two paths of length
           at least two leading from two of the vertices into distinct
           blocks, or with an edge leading from one of the three ver-
           tices into a different block, if the triangle is imbedded
           in a complete graph on four points,
  (iii) G contains no G_1 (two or all three peripheral points of
          G_1 may coincide).

<u>Proof</u>. The necessity of (i) and (iii) is clear. As to (ii), suppose
G contains a triangle with three edges attached to it as specified
in the theorem, then these six edges clearly induce G_3 in $L(G)$.

If the triangle has two paths of length at least two attached to it,
then these seven edges are seen to induce a 5-cycle with two chords
and two outgoing edges opposite the two chords, which is a non PO-
graph. Finally it may be verified that the line graph of a complete
graph on 4 points plus an edge does not admit a partial ordering,
thus completing the proof of the necessity part.

Suppose, then, that G satisfies the conditions of the theorem,
and let C_k be a longest cycle in G.

<u>Case a</u>. $k \geq 6$, and k is even. By (i) and (iii), C_k contains no
chords, and again by (iii) there can only be star graphs attached to
C_k, centered at the vertices of C_k, and any two star graphs must
be point-disjoint, as (i) clearly implies. The line graph of such a
graph is seen to be a PO-graph.

<u>Case b</u>. $k = 4$. Assume first that there are no chords in C_4, then
(iii) readily provides the answer. Next assume there is exactly one
chord in C_4. Let $C_4 = \{a, b, c, d\}$ with the chord (a, c). Again
applying (iii), we conclude there can only be star graphs attached to
C_4, centered at the vertices of C_4. Now (ii) implies that we either
have star graphs centered at a and c, or at b and d. The line

graphs of such graphs are PO-graphs. Finally if we have both chords in C_4, we appeal to (ii) again, noting that the line graph of the complete graph on 4 points admits a partial ordering.

Case c. $k = 3$. By (i) two triangles cannot be in the same block unless they have an edge in common. The hypothesis $k = 3$ and (ii) and (iii) now imply that there are at most two triangles, joined by a path (possibly of length zero, in which case, the triangles have a common vertex). Then (ii) and (iii) applied to the triangles and the connecting path settle this case.

Case d. There are no cycles. Making use of (iii) once more, we recognize G to be a path with point-disjoint star graphs attached to it, the line graph of which is a PO-graph.

Going through the cases of the proof of Theorem 13, it is easy to check that all those graphs G, whose associated line graphs L(G) are PO-graphs, are PO-graphs themselves. We state this as

Corollary 3. If G is a non PO-graph, then L(G) is a non PO-graph.

It is worth noting that Corollary 3 can also be verified by showing that a GH-cycle in G gives rise to a GH-cycle in L(G) (for details, see [1]).

For indifference systems, Lemma 7 immediately yields the next result.

Theorem 14. Let G be a graph, then L(G) is an I-graph if and only if G contains no G_1, G_2 or C_k for $k \geq 4$.

Corollary 4. If L(G) is an I-graph, then G is an I-graph or it contains K(1,3)'s as full subgraphs with at most two of the peripheral points having degree ≥ 2.

Theorem 15. Let G be a graph, then L(G) is a SO-graph if and only if G contains no $K_3 + K_2$, $K(1,3) + K_2$, G_1, G_2, $K(2,1) + K(2,1)$, C_5 or $K(3,2)$.

Proof. In light of Theorem 3 and Lemma 7, G must not contain G_1 or G_2 as subgraphs. Next, since L(G) cannot have $K_3 + K_1$ as a subgraph, we infer the impossibility of $K_3 + K_2$ or $K(1,3) + K_2$ in G. As to C_k ($k \geq 4$), we observe that the complements of C_4 and C_5 (both without chords) are $K_2 + K_2$ and C_5 (without chords), respectively. This now implies the nonexistence of $K(2,1) + K(2,1)$ and C_5 in G. Further, the line graph of $K(3,2)$ equals the complement of C_6 (without chords). Finally, using the theorem of [9] mentioned before, it is verified that L(G) cannot contain the complement of C_k (without chords) for $k \geq 7$.

Corollary 5. Let G be connected and let L(G) be a SO-graph. Then G is a SO-graph or the graph of Figure 6. (The dotted line means that this line plus the outer point may or may not be in the graph.)

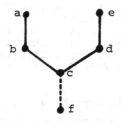

Figure 6

Proof. It follows from Theorem 16 and the connectivity of G that T fails to be a SO-graph only if it contains $K_2 + K_2$ as a full subgraph. Let (a,b) and (d,e) be two such edges. The impossibility of $K(2,1) + K(2,1)$ now implies that a path in G can have length at most 4. Considering the path linking (a,b) and (d,e) and calling the connecting vertex c, we infer that there can be no other edges incident with a, b, d or e (no $K(1,3) + K_2$!). As to c, only pendant edges can be attached to it (other than (b,c) and (c,d)), and again using the absence of $K(1,3) + K_2$, we conclude that there can be at most one such edge.

Theorem 16. Let G be a graph, then L(G) is an IV-graph if and only if G contains no C_k (k ≥ 4), G_1 or G_2.

Proof. According to Lemma 7, a line graph cannot contain $K(1,3)$ as a full subgraph. But now Theorem 4 shows that L(G) is an IV-graph if and only if it is an I-graph, and hence Theorem 14 applies.

An alternative proof is possible using Lemma 6 of Section III.

Corollary 6. If G is a non IV-graph, then L(G) is a non IV-graph.

Proof. According to Corollary 4, L(G) being an IV-graph implies that G is an I-graph (and, a fortiori, an IV-graph) or there are $K(1,3)$'s in G. The condition in Corollary 4 that every such $K(1,3)$ has at most two peripheral points of degree ≥ 2 now implies that G is of the form described in Lemma 6, and hence an IV-graph.

V. Uniqueness. If the binary relation R is symmetric, then G(R) obviously completely characterizes R. But if R is not symmetric, then not only do we have $G(R) = G(R^I)$, but there may be other relations R' with G(R') = G(R). Let us call a graph G uniquely partially orderable (UPO-graph) or uniquely semi-orderable (USO-graph) if G is a PO-graph or a SO-graph, and if R and R' are two relations such that G = G(R) = G(R'), then R' = R or $R' = R^I$.

In this section, we shall derive a characterization of USO-graphs and a sufficient condition for a PO-graph to be uniquely partially orderable. We will give all the necessary lemmas and definitions, but the proofs shall only be outlined. For the details, the reader is referred to [2].

Let us recall the algorithm designed by Gilmore and Hoffman in their characterization of PO-graphs. They show that one arrives at a partial ordering by assigning a direction to an arbitrary edge, orienting all edges whose direction is determined by previously oriented edges, choosing again an arbitrary undirected edge, etc. Since the direction of an edge may be forced in two possible ways, we give the following definitions.

Definition 7. The generalized path $a_1, \ldots, a_k$ is called a strong path from (a_1, a_2) if it does not contain triangular chords. The direction of the edge (c,d) is said to be strongly forced by the direction of (a,b) if there exists a strong path $a_1, a_2, \ldots, a_{k-1}, a_k$ with $(a,b) = (a_1, a_2)$, $(c,d) = (a_{k-1}, a_k)$.

Definition 8. $(a \rightarrow b)$ means the edge (a,b) has the orientation $(a \rightarrow b)$. The direction of the edge (a,c) is said to be transitively forced by the directions of (a,b) and (b,c), if either $(a \rightarrow b)$, $(b \rightarrow c)$ or $(b \rightarrow a)$, $(c \rightarrow b)$.

Definition 9. $G(a,b)$ denotes the set of all edges whose direction is forced by an initial orientation of (a,b) and subsequent strong and transitive forcings.

Lemma 8. If $(c,d) \in G(a,b)$, then $G(c,d) = G(a,b)$.

Proof. The hypothesis clearly implies $G(c,d) \subset G(a,b)$. If $(a,b) \in G(c,d)$ then the inverse also holds, and we are done. So suppose $(a,b) \notin G(c,d)$, and assume that $(a \rightarrow b)$ implies $(c \rightarrow d)$. Starting the Gilmore-Hoffman algorithm with the edge (c,d) and choosing the direction $(d \rightarrow c)$ we may assign $(a \rightarrow b)$ to (a,b), thus contradicting the hypothesis.

Lemma 8 gives a partitioning of $E(G)$ into orientation-blocks where the number of blocks represents the degree of freedom we have in constructing our partial order. Hence we have the following

Corollary 7. Let G be a PO-graph with m orientation-blocks. Then the number of partial orders R with $G(R) = G$ is 2^m. G is a UPO-graph if and only if $m = 1$.

We remark that some or all of these partial orders may be isomorphic.

Lemma 9. In an orientation-block $G(a,b)$, the direction of any edge strongly forces the direction of every other edge. In other words, transitive forcing can be replaced by strong forcing.

Proof. Strong forcing is easily seen to be a symmetric and transitive property, hence we are able to partition $G(a,b)$ into classes H_i with two edges being in the same class if and only if their directions force each other strongly. We then wish to show there is only one such class H_1. Assume the opposite and orient the edges of H_1. Among the edges in $G(a,b) - H_1$ there must be one whose direction is transitively forced by the directions of two edges in H_1. Let $(c,e) \in H_2$ be such an edge and assume that the two corresponding edges (c,d) and $(d,e) \in H_1$ have been directed $(c \to d)$ and $(d \to e)$.

Given this situation, it can be shown that every edge of H_2 is transitively forced by a pair of edges of H_1 with d as common endpoint But this implies that no two edges in H_2 with a common endpoint can receive transitive orientations. Hence, if we orient $G(a,b)$ by directing the edges of H_2 first, no transitive forcings would be possible, and $G(a,b) = H_2$, a contradiction.

Corollary 8. Let G be connected and contain no triangles, then G is a UPO-graph if and only if it is a PO-graph.

Proof. Every two edges are on a path, every path is strong, and hence Lemma 9 applies.

A trivial observation about UPO- and USO-graphs is that they must be connected. (They could have isolated points, but we agreed in Section II to rule those out.) Next, we give a result on the complement $C(G)$ which also applies to both classes of graphs.

Lemma 10. Let G be a UPO- or USO-graph and assume $C(G)$ is not connected, then $G = K(m,n)$ for suitable m,n.

Proof. Assume first that $C(G)$ possesses more than two components and let the vertex-sets of these components be $A_1, \ldots, A_n$. We orient the edges of $[A_i]$ in G arbitrarily (e.g., the orders induced by, respectively, a partial or semi-ordering of G). Every point of A_i is adjacent to every point of A_j, $j \neq i$. We orient these edges as follows: for $i < j$, $i \leq n-2$, orient the edges from $[A_i]$ to $[A_j]$. This yields two different orderings by directing the remaining edges (i) from $[A_{n-1}]$ to $[A_n]$ and (ii) from $[A_n]$ to $[A_{n-1}]$. It is easy to check that both orderings are well-defined. If $C(G)$ has two components, we follow the same procedure and obtain two different orderings as long as one of A_1 or A_2 is not completely disconnected in G. Finally, we remark that $K(m,n)$ is clearly both a UPO-graph and a USO-graph, and the proof is complete.

In order to give a characterization of USO-graphs, we need the next two results of [13].

Lemma 11. Let P be a semi-order and suppose $G = G(P)$. Then there exists a defining function f for the indifference system determined by $C(G)$ such that $a P b \Rightarrow f(a) < f(b)$ for all $a, b \in V(G)$.

Lemma 12. Let (A,I) be an indifference system and let G(I) be connected. Define a, b ∈ A to be equivalent if (i) a I b and (ii) a I c if and only if b I c for c ≠ a, b. Then if f and g are two defining functions for (A,I), and if there exists a pair of non-equivalent elements a, b ∈ A such that f(a) < f(b) and g(a) < g(b), then for every two non-equivalent elements x, y ∈ A we have f(x) < f(y) if and only if g(x) < g(y).

Lemmas 10, 11, and 12 can now be shown to imply the following characterization of USO-graphs.

Theorem 17. A SO-graph is an USO-graph if and only if it is either isomorphic to K(m,n) for some m and n, or if its complement is connected.

Corollary 9. A sufficient condition for a PO-graph to be uniquely partially orderable is that its complement be connected and contain neither a C_4 nor a K(1,3) as a full subgraph.

This follows immediately from Corollary 1 and the preceding theorem.

We wish to remove the restriction that C(G) does not contain a C_4 as a full subgraph. But then we may obtain graphs G with more than one component. If on the other hand, we confine ourselves to connected graphs, then we show that we can indeed omit the C_4 condition.

We state the following three lemmas whose proofs can be found in [2]. Also, for the rest of the paper, "subgraph contained in a graph" will mean a "full subgraph".

Lemma 13. Let G be a PO-graph and suppose C(G) contains no K(1,3). Let [a, b, c, d] be an arbitrary C_4 in C(G). If we delete the edge (a,c) (or (b,d)) from G, then the resulting graph G' is again a PO-graph and no K(1,3) is contained in its complement.

Lemma 14. Let G be a PO-graph and suppose C(G) contains no K(1,3). Let [a, b, c, d] be a C_4 in C(G), and assume G' = G − (a,c) is a UPO-graph. Then the reinsertion of (a,c) does not destroy the UPO-property of G'.

Lemma 15. Let G be a connected PO-graph whose complement C(G) is connected and does not contain a K(1,3). Let [a, b, c, d] be a C_4 in C(G) and assume G' = G − (a,c) is a UPO-graph. Then after reinsertion, (a,c) receives a unique orientation, i.e., G is a UPO-graph.

With this preparation, we are now in a position to prove

Theorem 18. Let G be a connected PO-graph whose complement is connected and contains no K(1,3), then G is a UPO-graph.

Proof. We proceed by induction on the number of edges in G, assuming the number of vertices p to be fixed throughout. The minimum possible number of edges is p-1, in which case G is a tree and hence a UPO-graph. Assume then the theorem holds for all graphs G' with $|V(G')| = p$, and $|E(G')| \leq e$. Let G be a graph which satisfies the hypotheses of the theorem and $|E(G)| = e+1$. If there are no C_4's in C(G), we are finished by Corollary 9, so let us assume the contrary. Two possibilities arise:

a. There exists at least one $C_4 = [a, b, c, d]$ in C(G) such that at least one of the edges (a,c), (b,d) is not a bridge in G.

b. There is no such C_4 in C(G).

<u>Case a</u>. Suppose that (a,c) is not a bridge in G. We delete (a,c) and invoking Lemma 13, we note that G' = G - (a,c) satisfies the induction hypothesis. Lemmas 14 and 15 now settle this case.

<u>Case b</u>. Let [a, b, c, d] be a C_4 in C(G). Assume that we are given the following structure of G as displayed in Figure 7, where the only edges between A, K, B are (a,c) and (b,d).

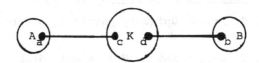

Figure 7

If there are no triangles in G, we are finished (Corollary 8). If there were a triangle in A (or B), then the three points making up the triangle plus b (or a) would induce a K(1,3) in C(G). Finally, a triangle in K contains at most one of the two vertices c, d, hence the triangle plus b or a again induces a K(1,3) in C(G), contradicting the hypothesis on G.

Using Lemma 9 and the foregoing theorem, we can extend Theorem 18 and obtain

<u>Corollary 10</u>. Let G be a connected PO-graph and let every triangle of G be contained in a connected full subgraph H of G, such that C(H) (i.e., the complement with respect to itself) is connected and contains no K(1,3). Then G is a UPO-graph.

We conclude with an example of a UPO-graph which is covered by Corollary 10, but not by the main theorem (figure 8).

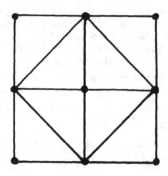

Figure 8

REFERENCES

1. M. Aigner, Graphs and partial orderings, <u>University of North Carolina Institute of Statistics Mimeo Series</u> No. 600.1.

2. M. Aigner and G. Prins, Uniquely partially orderable graphs, (to appear).

3. G. Chartrand, D. Geller and S. Hedetniemi, Graphs with forbidden subgraphs, <u>J. Combinatorial Theory</u> (to appear).

4. T. Gallai, Transitiv orientierbare Graphen, <u>Acta Math. Acad. Sci. Hungar</u>. 18 (1967), 25-66.

5. A. Ghouilá-Houri, Caractérisation des graphes nonorientés dont on peut orienter les aretes de maniére á obtenir le graphe d'une relation d'ordre, <u>C. R. Acad. Sci. Paris</u> 254 (1962), 1370-1371.

6. A. Ghouilá-Houri, Flots et tensions dans un graphe, <u>Ann. Scient. Éc. Norm. Sup</u>. 3e s.81 (1964), 207-265.

7. P.C. Gilmore and A.J. Hoffman, A characterization of comparability graphs and of interval graphs, <u>Canad. J. Math</u>. 16 (1964), 539-548.

8. B. Grünbaum, Incidence patterns of graphs and complexes, (this volume).

9. J. Krausz, Démonstration nouvelle d'un théoreme de Whitney sur les réseaux, <u>Math. Fiz. Lapok</u> 50 (1943), 75-85.

10. C.G. Lekkerkerker and J.C. Boland, Representation of a finite graph by a set of intervals on the real line, <u>Fund. Math</u>. 51 (1962), 45-64.

11. R.D. Luce, Semiorders and a theory of utility discrimination, <u>Econometrica</u> 24 (1956), 178-191.

12. J.A. Mitchem, A Characterization of Hypo-Outerplanar Graphs, Abstract 68T-H3, <u>Notices Amer. Math. Soc</u>. 15 (1968), 813.

13. F.S. Roberts, <u>Representations of indifference relations</u>, Ph.D. Thesis, Stanford University, 1968.

14. H. Sachs, On a problem of Shannon concerning perfect graphs and two more results on finite graphs, <u>Waterloo Conference on Combinatorial Mathematics</u> (1968).

15. D. Scott and P. Suppes, Foundational aspects of theories of measurement, <u>J. Symbolic Logic</u> 23 (1958), 113-128.

16. J. Sedlaček, Some properties of interchange graphs, <u>Proc. Symp</u>. <u>Smolenice</u> (1963), 145-150.

17. K. Wagner, Fastplättbare Graphen, <u>J. Combinatorial Theory</u> 3 (1967), 326-365.

GRAPH THEORY AND FINITE PROJECTIVE PLANES

Sabra S. Anderson, Eastern Michigan University

The combinatorial topics of finite projective planes, symmetric block designs, orthogonal latin squares, and difference sets are very closely interrelated and share many unsolved problems. It is the purpose of this paper to give a graphical description of finite projective planes and to call the attention of graph theorists to the major unsolved problems in this area, in the hope that research in this area may be expanded.

A (finite) <u>projective plane</u> π is a (finite) set of objects called points and a set of collections of points, called lines, which satisfy the following three axioms.
A1. Two distinct points of π lie on exactly one line.
A2. Two distinct lines of π intersect in exactly one point.
A3. There are four points, no three of which lie on the same line.
It can easily be shown that if one point of π lies on $n+1$ lines, then every point of π lies on $n+1$ lines; the <u>order</u> of π is defined to be the number n. A projective plane π_n of order n has n^2+n+1 points and n^2+n+1 lines.

The projective plane π_2 is shown in Figure 1 in the traditional manner; the "lines" in this graph are $(n+1)$-tuples rather than the 2-tuple concept of a line in graph theory. It is more useful, in a graph-theoretic analysis, to represent a projective plane π as a bipartite graph whose point set includes both the points and lines of π, with incidence in π inducing the lines of the graph. This graph $G(\pi)$ is regular of degree $n-1$ and has $2(n^2+n+1)$ points and $(n+1)(n^2+n+1)$ lines. $G(\pi_2)$ is shown in Figure 2.

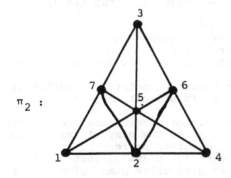

π_2 :

lines
$L_1 = (1,2,4)$
$L_2 = (2,3,5)$
$L_3 = (3,4,6)$
$L_4 = (4,5,7)$
$L_5 = (5,6,1)$
$L_6 = (6,7,2)$
$L_7 = (7,1,3)$

Figure 1

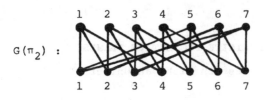

$$G(\pi_2) :$$

points of π

1 2 3 4 5 6 7

lines of π

Figure 2

The first unsolved problem is that of finding a characterization of those graphs G which are the graphs of projective planes. Hoffman [3] has found a characterization in terms of the eigenvalues of the adjacency matrix of G, but it does not distinguish between different planes of the same order. No characterization in terms of graphical properties exists.

<u>Theorem 1</u>. (Hoffman). Let G be a connected graph, regular of degree n+1, with $2(n^2+n+1)$ points. Then the adjacency matrix of G has the distinct eigenvalues n+1, -(n+1), $\sqrt{n}$, $-\sqrt{n}$ if and only if G is the graph of a projective plane of order n.

Somewhat greater success has been obtained in characterizing the line graph $L(\pi)$ of the graph of a projective plane π. Hoffman [3] has given a characterization in terms of the eigenvalues of its adjacency matrix, and Laskar and Dowling [4] have found a graphical description. Again, these characterizations do not distinguish between different projective planes of the same order.

Given two points u and v in a graph G, $\Delta(u,v)$ is defined to be the number of points in G adjacent to both u and v. If u and v are themselves adjacent, then $\Delta(u,v)$ is the <u>degree</u> of (u,v). A graph is <u>line regular</u> of degree k if every line has degree k.

<u>Theorem 2</u>. (Laskar and Dowling) The graph G is the line graph of a projective plane of order n if and only if it has the following properties:
1) G has $(n+1)(n^2+n+1)$ points.
2) G is connected and regular of degree 2n.
3) G is line regular of degree n-1.
4) $\Delta(u,v) = 1$ if the distance d(u,v) between u and v is 2.
5) If d(u,v) = 2, then the number of points w such that d(u,w) = 1 and d(v,w) = 2 is at most n-1.

The major unsolved problem is that of determining those values n for which a projective plane exists and how many different projective planes of order n exist. A survey of results in this area shows that there is exactly one projective plane of each order n = 2,3,4,5,7, and 8, there is no projective plane of order 6, and there is more than one projective plane of order 9. There is a projective plane every order $n = p^{\alpha}$, where p is a prime. The Bruck-Ryser Theorem [1] proves that there is no projective plane of order n when $n \equiv 1$ or 2 (mod 4) and there is a prime

$p \equiv 3 \pmod 4$ which divides the squarefree part of n.

The first undecided case is $n = 10$, and no projective plane whose order is a product of two or more distinct primes is known to exist. The Chowla-Ryser Theorem [2] states that whenever a projective plane of order n exists, then the Diophantine equation

$$z^2 = nx^2 + (-1)^{\frac{n^2+n}{2}} y^2$$

has a nontrivial solution. If the converse to this theorem is true, then there is a projective plane of order 10.

Now restricting ourselves to a special type of projective plane, we note that the projective planes π_2 (shown in Figure 1) and π_3 have their points labeled in such a way that the points of each line L_i can be obtained by adding 1 to the points of L_{i-1}. Such projective planes are called <u>cyclic</u>. A more concise definition states that the points and lines of a cyclic projective plane π can be labeled in such a way that the point-line incidence matrix of π is a circulant. For example, the point-line incidence matrix of π_2, labeled as in Figure 1, is

$$\begin{bmatrix} 1 & 1 & 0 & 1 & 0 & 0 & 0 \\ 0 & 1 & 1 & 0 & 1 & 0 & 0 \\ 0 & 0 & 1 & 1 & 0 & 1 & 0 \\ 0 & 0 & 0 & 1 & 1 & 0 & 1 \\ 1 & 0 & 0 & 0 & 1 & 1 & 0 \\ 0 & 1 & 0 & 0 & 0 & 1 & 1 \\ 1 & 0 & 1 & 0 & 0 & 0 & 1 \end{bmatrix} .$$

Cyclic projective planes have mainly been studied under the guise of difference sets. No graphical characterization has been found. There are cyclic projective planes of every order p^α, where p is a prime, (thus $\pi_2, \pi_3, \pi_4, \pi_5, \pi_7$, and π_8 are all cyclic), and there is a noncyclic projective plane of order 9. It has been conjectured that there is a cyclic projective plane of every order for which there is a projective plane. However, this conjecture and the converse to the Chowla-Ryser Theorem cannot both be true, since there is no cyclic projective plane of order 10.

This outlines the basic combinatorial features of finite projective planes and summarizes the current status of research on the major unsolved problems. It is hoped that a graph-theoretic approach to these problems may yield some new insights.

REFERENCES

1. R.H. Bruck and H.J. Ryser, The nonexistence of certain finite projective planes, <u>Canad. J. Math.</u> 1 (1949), 88-93.

2. S. Chowla and H.J. Ryser, Combinatorial problems, <u>Canad. J. Math.</u> 2 (1950), 93-99.

3. A.J. Hoffman, On the line graph of a projective plane, <u>Proc. Amer. Math. Soc.</u> 16 (1965), 297-302.

4. R. Laskar and T.A. Dowling, A geometric characterization of the line graph of a projective plane, <u>Institute of Statistics Mimeo Series</u> No. 516, University of North Carolina at Chapel Hill, 1967.

ON STEINITZ'S THEOREM CONCERNING CONVEX 3-POLYTOPES
AND ON SOME PROPERTIES OF PLANAR GRAPHS [1]

David W. Barnette, University of California at Davis
Branko Grünbaum, University of Washington

1. <u>Introduction</u>. The vertices and edges of every 3-polytope (i.e., 3-dimensional convex polytope) P determine a graph G(P), the <u>graph of</u> P. It is very easy to prove that G(P) is planar and 3-connected for each 3-polytope P. The converse statement constitutes the non-trivial part of the result known as

<u>Steinitz's Theorem</u>. A graph C is isomorphic to the graph G(P) of some 3-polytope P if and only if C is planar and 3-connected.

The main aim of this paper is to give two proofs for the non-trivial part of Steinitz's theorem. The first of our proofs follows one of Steinitz's own proofs (Steinitz [14], Steinitz-Rademacher [15]), the innovation being in the way two key lemmas are established. Our second proof is based on an idea of Kirkman [7] which was independently found by Tutte [17]. Both proofs use in part the same considerations (see Sections 2 and 4); the main difference between them is in the properties of 3-connected graphs they utilize (Sections 3 and 5).

2. <u>Two proofs of Steinitz's theorem</u>. The general ideas involved in both our proofs, as well as in the proof of Steinitz's theorem given in Grünbaum [2], may be stated as follows.

The proof is by induction on the number of edges of the graph C, the assertion being obvious if C is K_4, the complete graph with 4 vertices, which is the 3-connected graph having the smallest possible number of edges. In the inductive step, there are two stages:

I. We find a suitable method of associating with each 3-connected planar graph C having more than 6 edges another 3-connected planar graph C' which has fewer edges than C.

II. From any given 3-polytope P' such that G(P') is isomorphic to C' (the existence of such P' follows from the inductive assumption) we construct a 3-polytope P such that G(P) is isomorphic to C.

Clearly, the construction of P from P' in stage II depends on the manner in which C' was associated with C in stage I.

Now we proceed to the details of the two proofs.

---

[1] Research supported in part by National Science Foundation Grant GP-3470 and by Office of Naval Research contract N 00014-67-A-0103-0003.

<u>First Proof</u>. We shall say that a graph C' is obtained by <u>omitting an edge</u> E from a graph C provided the vertices of C and C' coincide, and C' contains all the edges of C except E (and no other edges). We shall say that C' is obtained from C by <u>deleting</u> an edge E provided C' results from C by omitting E, and if an endpoint of E would become of degree 2, by amalgamating the two edges incident to such a vertex into a single edge.

Since a graph contains no 1- or 2-circuits, not every edge of a graph is deletable (though each may be omitted). For a 3-connected planar graph C, the deletion of an edge E may take one of the three forms shown in Figure 1, depending on the number of vertices of degree 3 which are endpoints of E. For a 3-connected planar graph C, it is possible in case (1) that a_1 and a_2, or b_1 and b_2, coincide but both coincidences cannot happen simultaneously; similarly, a_1 and b_2, or b_1 and a_2, may coincide, but not both. In cases (2) and (3), the two coincidences of each pair may take place simultaneously.

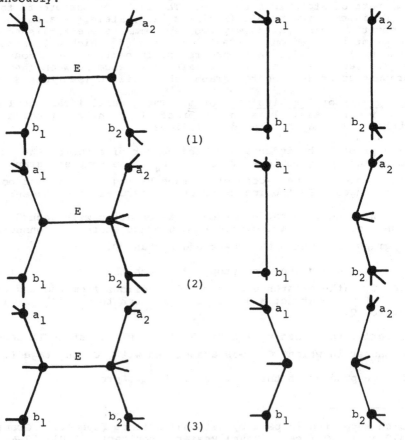

Figure 1.

In order to perform stage I of the proof we need

<u>Lemma 1</u>. Each 3-connected planar graph C with more than 6 edges
has an edge E such that the graph C' obtained from C by deleting
E is planar and 3-connected.

We defer the proof of Lemma 1 to Section 3 where it will be obtained as a corollary of a more general result which is of independent interest.

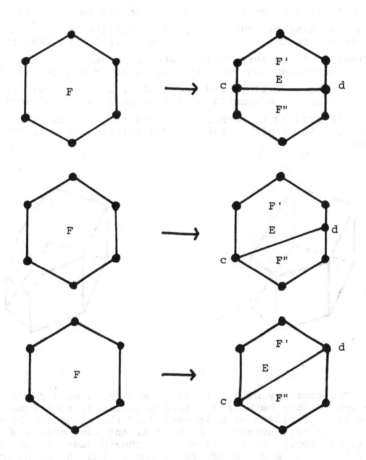

Figure 2

In order to perform stage II of our proof, we have to show how
to "invert", <u>on a given</u> 3-<u>polytope</u> P', the operation of deleting an
edge. In other words, we are given a 3-polytope P' with its graph

isomorphic to C', and a chord E = (c,d) on one of the faces F
of P' (see Figure 2) which "splits" F into two convex polygons
F' and F''. Both, one, or none of the endpoints c,d of E may be
among the vertices of F (and P'). The graph defined by the vertices and edges of P' together with E and its endpoints is isomorphic to C; unfortunately, this graph is not the graph of a 3-
polytope since F' and F'' are coplanar. Our aim being to find a
3-polytope with graph isomorphic to C the remedy would seem to lie
in turning F'' slightly about E (see the illustration in Figure
3). However, if some vertex of P' belonging to F'' but not to
F' has degree greater than 3, any such rotation of F'' induces
radical changes in the structure of the graph of the resulting polytope. Clearly, if we intend to rotate F'' about E, we must be
prepared to accompany this movement by suitable changes in the posi-
tions of the other vertices and faces of P'. It is far from obvious
that a collection of movements of the various faces and vertices of
P', suitably attuned to each other so as to yield the incidences required by C, is at all possible. The solution of this impasse results from

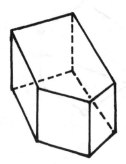 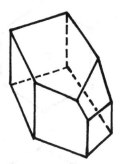

Figure 3

Lemma 2. The totality of all vertices and faces of a 3-connected
planar graph C may be arranged in a sequence in such a manner that
each element is incident with at most 3 of the elements which precede
it in the sequence. Moreover, such an arrangement exists even under
the additional requirement that for any chosen edge E of C the
two vertices of E and the two faces containing E be placed at the
beginning of the sequence.

   In the statement of Lemma 2, two elements are called incident
provided one of them is a face and the other a vertex of that face.

   We defer the proof of Lemma 2 till Section 4. Assuming its val-
idity, stage II of the proof of Steinitz's theorem becomes easy. We

arrange the vertices and faces of C in a sequence of the type described in Lemma 2, with F', F'', c and d at the beginning of the
sequence; then we may move the vertices and faces of P' to an ac-
commodating position for each sufficiently small rotation of F''. If
the element (vertex or face) is incident with 3 preceding ones, its
position is uniquely determined. If it is incident with fewer than 3,
we may choose it freely subject to the 2, 1, or 0 restrictions imposed on it by the elements which precede it. Since a plane [point]
determined by three points [three intersecting planes] depends continuously on the points [planes], all the moved elements may be chosen
to be in an arbitrarily small neighborhood of the original elements of
P'. But sufficiently small displacements of faces and vertices do not
destroy the convexity of a 3-polytope; hence the construction of the
3-polytope P with graph G(P) isomorphic to C is possible starting from any P' and E, and the proof of Steinitz's theorem is completed.

Second proof. Following Tutte [17], let any graph isomorphic to the
graph of an n-sided pyramid be called an n-<u>wheel</u>. By its very definition, each n-wheel satisfies Steinitz's theorem. It is therefore only
a minor modification of the general outline of the proof if in stage I
we consider only graphs which are not wheels. The result we need is

<u>Lemma 3</u>. Each 3-connected planar graph C which is not an n-wheel,
for any n ≥ 3, contains either (a) an edge E' such that the omission of E' from C yields a 3-connected planar graph C'; or (b)
an edge E'' such that the omission of E'' from C and the identi-
fication of its two endpoints to a single vertex yields a 3-connected
planar graph C''.

It is obvious that the operation (a) is identical to the case (3)
of deleting an edge (Figure 1), while (b) is usually described as
shrinking the edge E'' to a point. Moreover, (b) is <u>dual</u> to (a) in
the following sense: C'' is the dual of the graph obtained by per-
forming the operation (a) on the dual C* of the graph C (at the
edge E''* of C* which corresponds to the edge E'' of C).

We will present a proof of Lemma 3 in Section 5, together with a
discussion of its ramifications.

In order to perform stage II in this proof, we have only to remark that in case (a) we may use the same argument as in the first
proof, while in case (b) one may pass from P' to the dual polytope
P'*, perform on it the corresponding operation (which, by the above
remark is of type (a)) and finally take the dual of the resulting 3-
polytope.

This completes the second proof of Steinitz's theorem.

<u>Remarks (1)</u>. Steinitz formulated his theorem in terms of 2-dimension-
al cell complexes rather than graphs. However, the ideas of our first

proof follow the Steinitz argumentation closely; the claim to simplic-
ity of reasoning in our proof is based on the way in which we prove
the key Lemmas 1 and 2 (see Sections 3 and 4). Steinitz's proof was
reproduced in Lyusternik [11]; unfortunately, Lyusternik's presentation is far from complete since he skips over the necessity to prove
our Lemma 1 or Steinitz's equivalent of it.

(2). The proof of Steinitz's theorem in Grünbaum [2] uses still
another method in stage I (and, correspondingly, in stage II). It
seems, however, that each of the methods has certain advantages (see
Section 6).

3. <u>Deleting edges from 3-connected graphs</u>. If C' is a graph obtained by deleting an edge E from a graph C, and if C' is non-
planar, then C is obviously non-planar. Hence Lemma 1 will be proved if we establish

<u>Theorem 1</u>. Each 3-connected graph C with more than 6 edges has an
edge E such that the graph C' obtained from C by deleting E is
3-connected.

Instead of proving Theorem 1 directly, we shall first establish a
stronger result on the structure of 3-connected graphs. For its formulation we need a few definitions.

Let $\Pi = \{\pi_1, \ldots, \pi_n\}$ be a family of (simple) paths in a graph
G. We associate with G and Π a multi-graph (that is, a graph
which possibly has 1- or 2-circuits) G_Π in the following manner.

The vertices of G_Π are all those vertices of G which are endpoints of the paths π_i of Π, or belong to more than one π_i.
(Hence it is possible that a given π_i contains more than two vertices in G_Π.)

The edges of G_Π are the arcs of the π_i's determined (on the
π_i) by the vertices of G_Π.

We shall now prove the following result, from which Theorem 1
will be derived later.

<u>Theorem 2</u>. If G is a 3-connected graph, there exists a family
$\{\pi_0, \ldots, \pi_n\}$ of paths in G such that, for $\Pi(k) = \{\pi_0, \ldots, \pi_k\}$,
$1 \le k \le n$, we have:
  (i)   each of the multigraphs $G_{\Pi(k)}$, $1 \le k \le n$, is a graph and
        is 3-connected,
  (ii) $G_{\Pi(1)}$ is the graph K_4,
  (iii) $G_{\Pi(n)} = G$, and
  (iv) each π_{k+1} has only its endpoints in common with $G_{\Pi(k)}$,
        for $k = 1, \ldots, n-1$.

<u>Proof of Theorem 2</u>. For a given 3-connected graph G we shall induc-
tively define a family of paths π_i having the properties required by

the theorem. In order to find π_o and π_1, we need

Lemma 4. Every 3-connected graph C contains a system $T = \{\tau_1,\ldots,$ $\tau_6\}$ of paths such that C_T is K_4.

Proof of Lemma 4. Let x and y be two distinct vertices of C. Since C is 3-connected, there exist paths σ_1, σ_2, σ_3 in C, disjoint except for their common endpoints x and y. Since only one of the σ_i, say σ_1, is possibly an edge of C, σ_2 contains a vertex z of C different from x and y. Let Σ be the totality of paths in C having z as one of their endpoints. Some $\sigma \in \Sigma$ contains a point z' of $\sigma_1 \cup \sigma_3$ while $x,y \notin \sigma$, since otherwise the omission of x and y would disconnect C between z and some vertex in σ_3 different from x and y, in contradiction to the 3-connectedness of C. This σ contains a subpath σ_4 with endpoints v and w such that $v \in \sigma_2$, $w \in \sigma_1 \cup \sigma_3$, and $\sigma_4 \cap (\sigma_1 \cup \sigma_2 \cup \sigma_3) = \{v,w\}$. The six paths τ_i required for Lemma 4 are now easily found as the subpaths of σ_1, σ_2, σ_3 and σ_4 determined by x,y,v, and w. This completes the proof of Lemma 4.

Returning to the proof of Theorem 2, let T be a system of six paths in G of the type described in Lemma 4, such that G_T is the complete graph with vertices x,y,v,w. Among all the possible choices of T we choose one which involves the largest possible number of vertices of G. We define π_o as the union of the three paths of T with endpoints (x,y), (y,v), and (v,w), while π_1 is the union of the three other members of T. Then $G_{\Pi(1)}$ clearly satisfies the requirements of the theorem. If $G_{\Pi(1)} = G$, the proof of the theorem is completed. If not, we proceed by induction. Let $\pi_o,\ldots,\pi_k$ and hence $G_{\Pi(k)}$ be contructed; if $G_{\Pi(k)} = G$ we are done; otherwise, we distinguish two cases:

(A) There exists a vertex z of G which belongs to $G_{\Pi(k)}$ but is not a vertex of $G_{\Pi(k)}$;

(B) No such vertex of G exists, but there exists a vertex z of $G_{\Pi(k)}$ such that not all edges of G incident with z are in $G_{\Pi(k)}$.

Since we assumed $G_{\Pi(k)} \neq G$, one of those cases must occur.

In case (A), let the edge E of $G_{\Pi(k)}$ that contains z have endpoints z_1 and z_2, and let Σ be the totality of paths in G that start at z and contain neither z_1 nor z_2. By the 3-connectedness of G some $\sigma \in \Sigma$ has an endpoint z' which belongs to $G_{\Pi(k)}$ but not to the edge E. Hence there also exist paths in G which have one endpoint in the relative interior of E the other endpoint in $G_{\Pi(k)}$ but not in E, and do not meet $G_{\Pi(k)}$ except at their endpoints. We choose as π_{k+1} one such path which involves the largest possible number of vertices of G.

In case (B) all the edges of $G_{\Pi(k)}$ are also edges of G; then there exist paths in G which start at z and have only the endpoints in common with $G_{\Pi(k)}$. By the choice of each π_i, $0 \leq i \leq k$, as involving the maximal possible number of vertices of G, the endpoint different from z of each such path is a vertex z' of $G_{\Pi(k)}$ which is not a neighbor of z in $G_{\Pi(k)}$ (or in G). We choose as π_{k+1} one of the paths starting at z having those properties and involving the largest possible number of vertices of G.

In either case, the 3-connectedness of $G_{\Pi(k+1)}$ follows from that of $G_{\Pi(k)}$ and the well-known and easily established fact (see, for example, Harary [4], Theorem 3.3) that in every 2-connected graph H and for any two vertices (or two edges, or an edge and a vertex) of H there exists in H a circuit containing them.

The other assertions concerning π_{k+1} being obvious from the construction, this completes the proof of Theorem 2.

<u>Proof of Theorem 1</u>. Let $\{\pi_0, \pi_1, \ldots, \pi_n\}$ be a family of paths in C having the properties stated in Theorem 2. Since C is not the complete graph with 4 vertices we have $n \geq 2$. Since each vertex of C has degree at least 3, π_n consists of a single edge E of C. The graph that results from C on deleting E is isomorphic to $G_{\Pi(n-1)}$ and is therefore 3-connected. This completes the proof of Theorem 1, and thus also of Lemma 1.

<u>Remarks (1)</u>. While Theorem 2 seems to be completely new, many special cases of Theorem 1 and Lemma 1 have been considered in the literature. Thus Lemma 1 appears in Steinitz [14, p. 71] and in Steinitz-Rademacher [15, p. 192], though formulated for certain cell-decompositions of the 2-sphere. The special case of Lemma 1 dealing with 3-regular, 3-connected planar graphs is very easy to prove directly. We do not know by whom it was first established; it is mentioned in Steinitz-Rademacher [15, P. 228]. The 3-regular case of Theorem 1 (dealing with graphs which are not necessarily planar) was first established by Kotzig [9]; a somewhat weaker version was rediscovered by Johnson [5] (see also Ore [13, p. 129]).

(2). Lemma 4 seems to have been first proved by J. Isbell (see Grünbaum [2, p. 215]). For results extending Lemma 4 in various directions see Mader [12], Jung [6], and Larman-Mani [10].

4. <u>Proof of Lemma 2</u>. For any (3-connected) planar graph G we define a <u>bipartite</u> (3-connected) planar graph K(G) as follows:

The vertices of K(G) are in a one-to-one correspondence with the set of all vertices and faces of G. Two vertices of K(G) determine an edge of K(G) if and only if one of them corresponds to a face of G and the other to a vertex of that face.

The assertions that K(G) is planar and bipartite are trivial (see Figure 4, where G is drawn in dashed lines, K(G) in solid lines); the 3-connectedness of K(G) for 3-connected G is easy to establish but we do not need it.

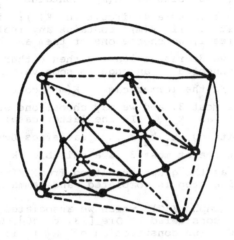

Figure 4

We recall the following well-known result, which is an immediate consequence of Euler's Theorem.

Lemma 5. Let H be a nonempty planar graph without isolated vertices, let $v_i(H)$ be the number of vertices of H having degree i, for $i \geq 1$, and let $p_j(H)$ be the number of faces of H incident (counting multiplicities) with j vertices of H. Then

$$\sum_i (4-i)v_i(H) + \sum_j (4-j)p_j(H) \geq 8.$$

We shall use the following corollary of Lemma 5.

Lemma 6. If H is a nonempty, planar, bipartite graph without isolated vertices, then

$$3v_1(H) + 2v_2(H) + v_3(H) \geq 8.$$

In order to prove Lemma 2, let k_1, k_2, k_3, k_4 be the four vertices of K(C) which correspond to the vertices and faces of C incident with the edge E of C. We shall construct a sequence $k_1, k_2, \ldots, k_n$ of all vertices of K(C), such that each k_j is connected by an edge of K(C) to at most three k_i's with $i < j$. This will clearly prove Lemma 2. To begin, we note that since no vertex of $K(C)$ has degree less than 3, $K(C)$ contains, by Lemma 6, at least eight 3-valent vertices. We choose one of them, different from k_1, k_2, k_3, k_4, and denote it by k_n. Let $K(n)$ be the graph obtained from $K(C)$ by omitting k_n and the edges incident with it.

Now we proceed by induction. If $k_n, k_{n-1}, \ldots, k_m$ are already determined together with $K(n), K(n-1), \ldots, K(m)$ and if $m > 5$, we shall define k_{m-1} as follows: $K(m)$ contains k_1, k_2, k_3, k_4, and these 4 vertices determine a 4-circuit in $K(m)$; in particular each has degree at least 2. If $K(m)$ contains any isolated vertices or any vertex of degree 1, we choose one of them as k_{m-1}. If all vertices of $K(m)$ have degree at least 2 then either at least one of k_1, k_2, k_3, k_4 is connected by an edge to some vertex not in $\{k_1, k_2, k_3, k_4\}$, or not. In the former case, $K(m)$ contains at least 5 vertices of degree at most 3, and we may choose one of the (different from k_1, k_2, k_3, k_4) as k_{m-1}. In the latter case, the graph obtained from $K(m)$ by omitting k_1, k_2, k_3, k_4 satisfies Lemma 6, and any of its vertices of degree at most 3 is a suitable k_{m-1}. Defining in any case $K(m-1)$ as the graph resulting from $K(m)$ by omitting k_{m-1}, the inductive step is completed, and Lemma 2 proved.

Remarks (1). The graph $K(C)$ which we associated with a planar graph C has been considered by Ore [13, p. 46], who calls it the radial graph of C. The construction of $K(C)$ is easily seen to be the dual of the θ-construction of Steinitz [14] and Steinitz-Rademacher [15, p. 196], considered also (under the name $I(C)$) in Grünbaum [2, p. 241] and (as the medial graph of C) in Ore [13, p. 47 and p. 124].

(2). With any d-polytope P it is possible to associate a bipartite graph $K(P)$ in the following fashion: The vertices of $K(P)$ correspond to the vertices and facets (i.e., maximal proper faces) of P, two vertices of $K(P)$ determining an edge if and only if one of them corresponds to a facet of P and the other to a vertex of that facet. It seems that there is some interest in investigating the properties of such graphs $K(P)$ and, in particular, in the following question: Is it always possible to order the vertices of $K(P)$ in a sequence in such a fashion that each is joined by an edge to at most $n(d)$ vertices preceding it, where $n(d)$ depends on the dimension d of P but not on the particular d-polytope P. Lemma 2 may be interpreted as saying that $n(3) = 3$. It may be conjectured that $n(d) < \infty$ for all d, though the question is open already for $d = 4$. The examale of the regular 24-cell in 4 dimensions shows at any rate that $n(4) \geq 6$.

For somewhat related questions see Grünbaum [3].

5. Proof of Lemma 3. We shall start by proving

Lemma 7. Every 3-connected planar graph which is not a wheel contains either an edge which is not the side of any triangle, or an edge which is not incident to any vertex of degree 3.

Proof of Lemma 7. If all vertices of G have degree 3 and all faces

are triangles, then G is K_4. Therefore, using duality, we may assume that some face of G is a polygon with at least 4 sides. Let
F be such a face; then there is nothing to prove unless all the
edges of F belong to triangles. For some edge E of F, at least
one endpoint V of E must have degree greater than 3 (see Figure
5a); indeed, if all vertices of F had degree 3, then G would be
a wheel. Then, in the notation of Fig. 5a, if either V_2 of V_3
have degree at least 4, we are done; hence we may assume both to be
of degree 3. By the above, the face different from F adjacent to
E' (see Figure 5b) is a triangle, since V_2 has degree 3, it follows that V_1 and V_4 belong to a face of G different from F,
i.e. G is not 3-connected. The contradiction reached completes the
proof of Lemma 7.

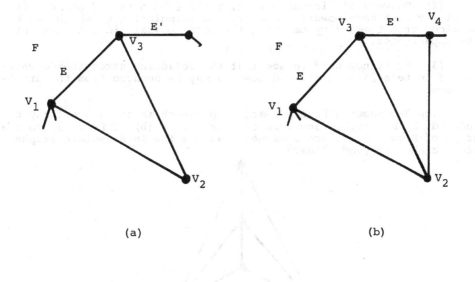

Figure 5

Proof of Lemma 3. Let E = (e,f) be an edge of C such that neither of the faces A and B adjacent to it is a triangle. (The existence of E follows from Lemma 7, and, possibly, duality.) If E
is not shrinkable, some faces F and G such that e ∈ F and
f ∈ G must have a common vertex g (possibly whole edge (g,h).).
We distinguish two cases.

Case 1. Both e and f have degree at least 4. Then E may be

omitted unless A and B are neighbors of one face H. But then H
would have to be either F or G, and then the 3-connectedness of
C would imply that e or f has degree 3.

Case 2. The vertex e has degree 3. Then the 3-connectedness of C
implies that F is not a triangle and that the edges A ∩ F = (e,a)
and B ∩ F = (e,b) are disjoint from G. If (e,b) cannot be
shrunk, e and b must belong to faces L,M different from B and
F such that e ∈ L, b ∈ M and L ∩ M ≠ ∅. But the only possibility
for L is L = A, and then by the Jordan theorem, A ∩ M ≠ ∅ implies that F ∩ G is a single point g which coincides with A ∩ M.
But then A ∩ F = (e,g), that is, a = p, which is a contradiction.

This completes the proof of Lemma 3.

Remarks (1). The ideas of Lemma 3 and of the above proof go back to
Kirkman [7]. However, as justly criticized by Steinitz [14, p. 72],
Kirkman was ostensibly working with vertices, edges, and faces of 3-
polytopes; hence to make his arguments valid, Steinitz's theorem (or
some other powerful tool) would have to be available first.

(2) Unaware of Kirkman's work, Tutte [17] obtained Lemma 3 as a
corollary of a more powerful result, the formulation of which may be
obtained from Lemma 3 by omitting the word "planar" in all three of
its appearances.

(3) It is not hard to see that the following strengthened version of Tutte's [17] result and Lemma 3 may be deduced from the original one:

Each 3-connected [and planar] graph which is not a wheel may be
reduced, by a suitable sequence of sets (a) or (b) of Lemma 3, to the
graph of Figure 6 in such a manner that all the intermediate graphs
are 3-connected [and planar].

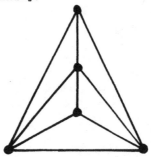

Figure 6

6. Additional remarks (1). The two proofs of Steinitz's theorem
presented here seem to have some advantages over the proof in Grünbaum [2]. They involve fewer "reductions" in Stage I, and the loca-
tion at which the reduction is to be performed is found in a simpler
fashion; this implies convenience in certain computational prodecures
(see the related remarks and references in Klee [8, p. 139]). On the
other hand, some results provable by easy extensions of the method in
Grünbaum [2] do not seem to be as easily accessible through the present proofs. As an example of this situation we mention the follow-

ing theorem (Grünbaum [2, p. 244]):

Every 3-polytope may be approximated arbitrarily closely by 3-polytopes of the same combinatorial type which have all vertices at points with rational coordinates (in any preassigned Cartesian system of coordinates.)

(2) As an obvious corollary to Steinitz's theorem one may prove the following result (which was established directly, and independently, by Wagner [18], Fáry [1], and Stojaković [16]: Every planar graph is isomorphic to a planar graph in which all the edges are straight-line segments.

REFERENCES

1. I. Fáry, On straight line representation of planar graphs, Acta Sci. Math. (Szeged), 11 (1948), 229-233.

2. B. Grünbaum, Convex polytopes, Wiley, New York, 1967.

3. B. Grünbaum, Higher-dimensional analogues of the four-color problem and some inequalities for simplicial complexes, J. Combinatorial Theory, to appear.

4. F. Harary, Graph theory, Addison-Wesley, Reading, 1969.

5. E.L. Johnson, A proof of 4-coloring the edges of a cubic graph, Amer. Math. Monthly, 73 (1966), 52-55.

6. H.A. Jung, Zusammenzüge und Unterteilungen von Graphen, Math. Nachr., 35 (1967), 241-267.

7. T.P. Kirkman, On autopolar polyhedra, Philos. Trans. Roy. Soc. London, 147 (1857), 183-215.

8. V. Klee, Convex polytopes and linear programming, Proc. IBM Scientific Computing Symposium on Combinatorial Problems, March 16-18, 1964, pp. 123-158 (1966).

9. A. Kotzig, O istých rozkladoch grafu. Mat.-fyz. casop. Sloven. Akad. Vied, Bratislava, 5 (1955), 144-151.

10. D.G. Larman and P. Mani, On the existence of certain configurations within graphs and the 1-skeletons of polytopes, to appear.

11. L.A. Lyusternik, Convex figures and polyhedra (in Russian), Moscow, 1956. (English translations: Heath and Co., Boston 1966, and Dover, New York 1963).

12. W. Mader, Homomorphieeigenschaften und mittlere Kantendichte von Graphen, Math. Ann., 174 (1967), 265-268.

13. O. Ore, The four color problem, Academic Press, New York 1967.

14. E. Steinitz, Polyeder und Raumeinteilungen, Enzykl. Math. Wiss., Vol. 3 (Geometrie), Part 3 AB 12, pp. 1-139 (1922).

15. E. Steinitz and H. Rademacher, _Vorlesungen über die Theorie der Polyeder_, Springer, Berlin 1934.

16. M. Stojaković, Über die Konstruktion der ebenen Graphen, _Univ. Beograd. Godisnjak Filozof. Fak. Novom Sadu_, 4 (1959), 375-378.

17. W.T. Tutte, A theory of 3-connected graphs, _Indag. Math._, 23 (1961), 441-455.

18. K. Wagner, Bemerkungen zem Vierfarbenproblem. _Jber. Deutsch. Math. Vereim._, 46 (1936), 26-32.

ANALOGUES OF RAMSEY NUMBERS[1]

Mehdi Behzad, Pahlavi University, Iran and
Western Michigan University

1. **Introduction**. The Ramsey number $f_1(r,s)$ is the smallest positive integer p such that every graph of order p contains r mutually adjacent or s mutually nonadjacent vertices. These numbers have been studied by, among others, Erdös [2,3,4 and 5], Graver and Yackel [6], Greenwood and Gleason [7], and Kalbfleisch [8]. Although various bounds for $f_1(r,s)$ have been given, the exact values remain unknown in general.

The edge analogue of this problem was considered in [1] by defining $f_2(r,s)$ as the smallest positive integer p such that every connected graph of order p contains r mutually adjacent or s mutually nonadjacent edges. It is essential to include the word "connected" in the above, for otherwise $f_2(r,s)$ is not well-defined. On the other hand, except for the case $r = 2$, $f_1(r,s)$ may have been defined by restricting to the class of connected graphs.

Another natural analogue of the Ramsey number in which both vertices and edges of graphs were used has been considered also. Two elements (vertices or edges) of a graph are <u>joint</u> if they are either adjacent (two vertices or two edges) or incident (a vertex and an edge); otherwise they are called <u>disjoint</u>. For positive integers r and s, we define $f_3(r,s)$ (respectively, $f_4(r,s)$) to be the smallest positive integer p such that every graph (connected graph) of order p has r mutually joint or s mutually disjoint elements.

In this note we present an outline of the proof of the main theorem of [1], and summarize a few results involving f_3 and f_4.

2. **The Edge Analogue of Ramsey Numbers**. The following theorem determines the value of $f_2(r,s)$ for all positive integers r and s.

Theorem 1. For $r = 2$ and $s > 1$ we have $f_2(2,s) = 3$; while $f_2(r,s) = (r - 1)(s - 1) + 2$ for all other positive integers r and s.

Since $f_2(r,1) = 2$ for every r, $f_2(1,s) = 2$ for every s, and $f_2(2,s) = 3$ for every $s > 1$, we confine ourselves to $r > 2$ and $s > 1$.

Lemma 1.1. If $f_2(r,s)$ is the smallest positive integer p such

---

[1]Research supported in part by a grant from the Office of Naval Research.

that every tree of order p contains r mutually adjacent or s
mutually nonadjacent edges, then $f_2(r,s) = f_2'(r,s)$.

The above lemma permits us to restrict ourselves to trees.
Moreover, the fact that every nontrivial tree contains two adjacent
vertices a and b with deg a = 1, such that at most one vertex
adjacent with b has degree greater than one, is used to prove the
following:

Lemma 1.2. For r > 2, s > 1, $f_2(r,s) > (r - s)(s - 1) + 1$.

Lemma 1.3. For r > 2, s > 1, $f_2(r,s) = (r - s)(s - 1) + 2$.

To prove this, define a class of trees, C(r,s), r > 2, s > 1,
of order $f_2(r,s) - 1$ containing neither r mutually adjacent nor
s mutually nonadjacent edges. Then use induction on s as follows.
For a fixed r > 2, assume that $f_2(r,s-1) = (r - 1)(s - 2) + 2$ and
that the members of C(r,s-1) are the only trees of order greater
than or equal to (r - 1)(s - 2) + 1 which contain neither r mutually adjacent nor s mutually nonadjacent edges.

The proof of Theorem 1 now follows.

3. The Vertex-Edge Analogue of Ramsey Numbers. The vertex-edge analogue of the Ramsey number for graphs is now considered. By induction on s, one first shows that $f_3(r,s) \le 2s - 1$ for all r and
s. Then the following is established.

Theorem 2. For all $s \ge 1$; $f_3(1,s) = 1$, $f_3(2,s) = f_3(3,s) = s$, and
$f_3(r,s) = 2s - 1$ for $r \ge 4$.

The number $f_4(r,s)$ can be determined for all lattice points
outside the region determined by r > 4 and $s > r^2 - 5r + 8$ in the
first quadrant of the (r,s)-plane. This is the next result.

Theorem 3. Let r and s be positive integers. Then

(1) $\qquad\qquad f_4(1,s) = f_4(r,1) = 1$ for all r and s,

(2) $\qquad\qquad f_4(2,s) = f_4(3,s) = 2$, $s \ge 2$,

(3) $\qquad\qquad f_4(r,2) = 3$, $r \ge 4$,

(4) $\qquad\qquad f_4(4,s) = [(3s + 1)/2]$, $s \ge 2$,

(5) $\qquad\qquad f_4(r,s) = 2s - 1$, $2 < s < r$, and

(6) $\qquad\qquad f_4(r,s) = 2s - 2$, $4 \le r \le s \le r^2 - 5r + 8$.

The first three statements of Theorem 3 are immediate. If a
connected graph does not contain four mutually joint elements, then
it is either a path or a cycle. This proves (4). Next, Theorem 2
together with the fact that the complete bipartite graph K(s - 1,
s - 1) contains neither r mutually joint nor s mutually disjoint
elements, 2 < s < r, yields (5).

In order to prove (6), we employ the following.

Lemma 3.1. For every positive integer $s (> 2)$, there exists exactly one connected graph of order $2s - 2$ containing neither $s + 1$ mutually joint nor s mutually disjoint elements.

Using Lemma 3.1, we next prove:

Lemma 3.2. For integers r and s, $4 \le r \le s$, $f_4(r,s) \le 2s - 2$.

Now let t be the greatest integer such that $t^2 - 5t + 8$ is less that s. The hypotheses of (6) implies that $4 \le t \le r - 1$. Observing that $s \le t^2 - 3t + 4$, we distinguish two cases:

(i) $t^2 - rt + 6 < s \le t^2 - 3t + 4$; and

(ii) $t^2 - 5t + 8 < s \le t^2 - 4t + 6$.

In either case, it is shown that $f_4(r,s) \ge 2s - 2$. This, together with Lemma 3.2 proves (6). This completes the proof of the Theorem.

REFERENCES

1. M. Behzad and H. Radjavi, The line analogue of Ramsey Numbers, Israel J. Math., 5 (1967), 93-96.

2. P. Erdös, Some remarks on the theory of graphs, Bull. Amer. Math. Soc., 53 (1947), 292-294.

3. P. Erdös, Remarks on a theorem of Ramsey, Bull. Res. Counc. Israel, 7F (1957), 21-24.

4. P. Erdös, Graph theory and probability I, Canad. J. Math., 11 (1959), 34-38.

5. P. Erdös, Graph theory and probability II, Canad. J. Math., 13 (1961), 346-352.

6. J.E. Graver and J. Yackel, An upper bound for Ramsey numbers, Bull. Amer. Math. Soc., 72 (1966), 1076-1079.

7. R.E. Greenwood and Gleason, Combinatorial relations and chromatic graphs, Canad. J. Math., 7 (1955), 1-7.

8. J. Kalbfleisch, On an unknown Ramsey number, Mich. Math. J., 13 (1966), 385-392.

The author's present address is the National University of Iran in Tehran.

A SURVEY OF PACKINGS AND COVERINGS OF GRAPHS

Lowell W. Beineke, Purdue University at Fort Wayne

The general covering problem in graph theory asks for the minimum number of graphs with a particular property having a given graph as their union. In contrast, a packing problem asks for the maximum number of edge-disjoint subgraphs having a specified property. These problems thus fit into the general category of packings and coverings and have generally dual natures--one asking for a minimization of the number of items needed for inclusion of a given object, the other for a maximization of the number of things to be included.

There are, however, few known cases in which solutions to packing and covering problems for graphs are related. Only one such relationship is given here, this for the problems of packing with spanning trees and of covering with forests. For problems other than these two, we concentrate on complete graphs and complete bipartite graphs.

The covering problems considered here use, in addition to forests, graphs with certain embeddability properties, including planar, planar bipartite, and toroidal graphs. Besides packings with spanning trees, we consider those using cycles, nonplanar graphs, nonouterplanar graphs, and regular complete bipartite graphs.

1. <u>Covering by forests</u>. An excellent example of a covering problem for an arbitrary graph G is the minimum number of forests whose union is G; this is sometimes called the <u>arboricity</u> of G. A solution to this problem has been found by Nash-Williams, [19], simply in terms of the number of edges in the induced subgraphs of G.

<u>Theorem 1</u>. The minimum number of forests needed to cover a graph G is

$$\max_{2 \leq n \leq p} \left\{ \frac{q_n}{n - 1} \right\}$$

where p is the number of vertices of G and q_n is the maximum number of edges in any subgraph of G with n vertices.

This theorem is exceptional in providing a solution for all graphs. In most covering and packing problems, only partial results or bounds have been obtained; and in such cases, attention is usually restricted to certain classes of graphs, in particular, the complete graphs K_p and the complete bipartite graphs $K(m,n)$. The arboricity solutions for these graphs are that K_p can be covered with $\left\lceil \frac{p + 1}{2} \right\rceil$ forests and $K(m,n)$ with $\left\{ \frac{mn}{m+n-1} \right\}$. Specific constructions realizing these numbers are given in [2].

The theorem of Nash-Williams has other interesting corollaries, one of which is that every planar graph is the union of three or fewer forests.

2. <u>Covering by planar graphs</u>. Some of the more intriguing covering problems involve subgraphs embeddable in certain surfaces. Of these, the most interesting case is of course the plane. The <u>(planar)</u> <u>thickness</u> of a graph G is the minimum number of planar subgraphs whose union is G. Even for the two standard classes of graphs (the complete and the complete bipartite) the problem remains partially unsolved.

The usual technique for attacking such a problem is to find an upper bound (for example, by Euler's polyhedron formula, as in this case) and then to attempt to find a constructive decomposition which attains this number. For thickness, there are two known exceptions to such a solution among complete graphs. The minimum number of planar graphs in a covering of K_9 and K_{10} was shown to be 3 by Battle, Harary, and Kodama [1], and a number of others. The following theorem summarizes the known cases in which the bound given by Euler's formula is correct.

<u>Theorem 2</u>. The minimum number of planar graphs required to cover K_p, for $p \neq 9$ or 10, is $\left[\frac{p+7}{6}\right]$, except possibly when $p \equiv 4 \pmod 6$. This number is also correct for $p = 4, 22, 28, 34,$ and 40.

The first statement of the theorem is due to Beineke and Harary [8]. However, each of the known cases of $p \equiv 4 \pmod 6$ has required a different construction: the case $p = 4$ is of course trivial, the case of 22 was shown by Hobbs and Grossman [16], 28 by Beineke [4], and 34 and 40 by Mayer [17]. It is of course conjectured that the number $\left[\frac{p+7}{6}\right]$ is correct for all higher values of p. The only smaller missing answer is for $p = 16$, and here the consensus is that the thickness is 5, primarily because so many unsuccessful attempts have been made.

The problem of planar coverings of complete bipartite graphs also has some gaps, but these do not occur very often nor in a known regular pattern. The theorem is due to Beineke, Harary, and Moon [9, 3].

<u>Theorem 3</u>. The minimum number of planar graphs required to cover $K(m,n)$ is

$$\left\{\frac{mn}{2(m+n-2)}\right\}$$

except possibly when $m < n$, m and n are both odd, and there is an integer k with $\frac{m+5}{4} \leq k \leq \frac{m-3}{2}$ for which

$$n = \left[\frac{2k(m-2)}{m-2k}\right].$$

For a given odd value of m, the number of possible exceptions with $m < n$ cannot therefore exceed $\frac{1}{4}(m-7)$. For $m \leq 30$, the total number of such unknown cases is only six: $K(19,29)$, $K(19,47)$, $K(23,79)$, $K(25,59)$, $K(27,71)$, $K(29,129)$. In contrast to the complete case, there is no known complete bipartite graph for which the equality fails to hold.

3. <u>Covering by planar bipartite graphs</u>. A variation of thickness, in which the planar subgraphs must also be bipartite, has been studied by Walther [22]. For complete graphs there is again an infinite class of unsolved cases.

<u>Theorem 4</u>. The minimum number of planar bipartite graphs needed to cover K_p, for $p \not\equiv 2 \pmod 4$, is $\left\lceil \dfrac{p+5}{4} \right\rceil$, except that K_5 requires 3 and K_9 requires 4.

It follows from the theorem that the bipartite thickness of K_2 is 1, of K_6 is 3, and of K_{10} is 4. All other cases of $p \equiv 2 \pmod 4$ are unknown.

4. <u>Covering by graphs embeddable in the torus, projective plane, and other surfaces</u>. Some cases of coverings of complete graphs with subgraphs embeddable in surfaces other than the plane have been investigated, and exact results have been found in three cases, the projective plane, the torus, and the double torus. These are due to Ringel [20] and Beineke [5].

<u>Theorem 5</u>. The minimum number of S-embeddable graphs required to cover K_p for $p \geq 3$ is

$\left\lceil \dfrac{p+5}{6} \right\rceil$ when S is the projective plane,

$\left\lceil \dfrac{p+4}{6} \right\rceil$ when S is the torus,

$\left\lceil \dfrac{p+3}{6} \right\rceil$ when S is the double-torus .

Franklin [12] has shown that K_7 cannot be embedded in the Klein bottle (non-orientable surface with two crosscuts). Thus, although that surface has the same Euler characteristic as the torus, the S-thickness results are not always the same.

For the general complete bipartite graph, present methods always leave exceptions, just as in the planar case (and in fact they are more numerous). In no case, however, is there a known exception in which the number obtained from the Euler pohyedron formula is incorrect, in contrast to the three known exceptions for complete graphs. For regular complete bipartite graphs $K(n,n)$ results are exact in a number of cases; see [5].

<u>Theorem 6</u>. The minimum number of S-embeddable graphs required to cover $K(n,n)$ for $n > 1$ is

$\left\lceil \dfrac{n+5}{4} \right\rceil$ when S is the plane,

$\left\lceil \dfrac{n+3}{4} \right\rceil$ when S is the torus,

$\left\lceil \dfrac{n+3}{4} \right\rceil$ when S is the double-torus,

$$\left\lceil \frac{n+2}{4} \right\rceil \quad \text{when} \quad S \quad \text{is the triple-torus,}$$

$$\left\lceil \frac{n+4}{4} \right\rceil \quad \text{when} \quad S \quad \text{is the projective plane,}$$

$$\left\lceil \frac{n+3}{4} \right\rceil \quad \text{when} \quad S \quad \text{is the Klein bottle.}$$

5. <u>Packing with spanning trees</u>. Packing graphs with spanning trees is closely related to covering graphs with forests, in that Nash-Williams [18] used similar methods in providing complete solutions to both problems. The general solution to the packing problem was first found by Tutte [21], however.

<u>Theorem 7</u>. The maximum number of spanning trees in a packing of a graph G is

$$\min_{2 \le k \le p} \left\lceil \frac{q_k}{k-1} \right\rceil ,$$

where p is the number of vertices of G and q_k is the maximum number of edges joining different sets in a partition of the vertices of G into k sets.

6. <u>Packing with cycles</u>. Generally speaking, what one would expect to be the packing problem dual to the problem of covering with graphs having with property P is packing with graphs not having property P. Thus, the problem dual to covering with forests would be packing with graphs which are not forests; that is, graphs with cycles. There is however no apparent connection between solutions of these two problems, in contrast to the relation of the packings with spanning trees to the coverings by forests.

   Packing with cycles have been found for the usual two families of graphs K_p and $K(m,n)$. The complete results are due to Chartrand, Geller, and Hedetniemi [10], but they use earlier results of Fort and Hedlund and of Guy for the complete case.

<u>Theorem 8</u>. The maximum number of cycles in a packing of K_p is $[\frac{p}{3}[\frac{p-1}{2}]]$.

<u>Theorem 9</u>. The maximum number of cycles in a packing of $K(m,n)$ with $m \le n$, is $[\frac{m}{2}][\frac{n}{2}]$ if mn is even and $[\frac{m}{2}][\frac{n}{2}] + [\frac{m}{4}]$ if mn is odd.

7. <u>Packing with nonplanar graphs</u>. A maximum packing with nonplanar graphs is in a sense opposite to thickness, just as a maximum packing with cycles is opposite to arboricity. This problem goes by the name of <u>coarseness</u>, and was first posed by Erdös, who asked whether K_{3r} can have more than the obvious $\binom{r}{2}$ edge-disjoint nonplanar subgraphs. By proving that it can, as given in the next theorem, Guy [13] had some of his research supported by Erdös. For both K_p and

K(m,n), the known results are incomplete, and due to Guy and Beineke [15,7]). The complete case is "eight-ninths" solved, and all are known to within 1.

<u>Theorem 10</u>. The maximum number $c(K_p)$ of nonplanar graphs in a packing of K_p satisfies the following equations and inequalities:

For $p = 3r$, $\quad c(K_p) = \binom{r}{2} \quad$ if $r \leq 5$

$$\binom{r}{2} \leq c(K_p) \leq \binom{r}{2} + 1 \quad \text{if} \quad 6 \leq r \leq 9$$

$$c(K_p) = \binom{r}{2} + \lceil \tfrac{r}{5} \rceil \quad \text{if} \quad r \geq 10$$

For $p = 3r + 1$, $\quad c(K_p) = \binom{r}{2} + \lceil \tfrac{2r}{3} \rceil$ if $r \not\equiv 2 \pmod 3$, $r \neq 3, 4$

$$c(K_{10}) = 4, \quad 7 \leq c(K_{13}) \leq 8$$

$$\binom{r}{2} + 2\lceil \tfrac{r}{3} \rceil \leq c(K_p) \leq \binom{r}{2} + \lceil \tfrac{2r}{3} \rceil \quad \text{if} \quad r \equiv 2 \pmod 3$$

For $p = 3r + 2$, $\quad c(K_p) = \binom{r}{2} + \lceil \tfrac{14r+1}{15} \rceil$.

<u>Theorem 11</u>. The maximum number $c(K(m,n))$ of nonplanar graphs in a packing of $K(m,n)$ satisfies the following equations and inequalities, when $m = 3p + d$, $n = 3q + e$, and $0 \leq d \leq e \leq 2$:

For $d = 0$ or 1 and $e = 0$ or 1, $\quad c(K(m,n)) = pq + \min (\lceil \tfrac{ep}{3} \rceil \lceil \tfrac{dq}{3} \rceil)$

For $d = 0$ and $e = 2$, $\quad\quad\quad\quad d(K(m,n)) = pq + \lceil \tfrac{p}{3} \rceil$, if $q > 0$.

For $d = 1$ and $e = 2$,
$\quad c(K(m,n)) \leq pq + \min (\lceil \tfrac{p+q}{3} \rceil, \lceil \tfrac{2p}{3} \rceil, \lceil \tfrac{16p+8q+2}{39} \rceil)$

$\quad c(K(m,n)) \geq pq + \max (\lceil \tfrac{p+2}{3} \rceil, \min (\lceil \tfrac{2p}{3} \rceil, \lceil \tfrac{q}{3} \rceil))$,
$$\text{if} \quad p \geq 2 \quad \text{and} \quad q \geq 7.$$

These are equal if $p \geq 2q$.

For $d = e = 2$,
$\quad c(K(m,n)) \leq pq + \min (\lceil \tfrac{p+2q}{3} \rceil, \lceil \tfrac{2p+q}{3} \rceil, \lceil \tfrac{16p+16q+4}{39} \rceil)$

$\quad c(K(m,n)) \geq pq + \lceil \tfrac{p}{3} \rceil + \lceil \tfrac{q}{3} \rceil + \lceil \tfrac{p}{9} \rceil$, if $0 < p \leq q$.

8. <u>Packing with non-outerplanar graphs</u>. Between the class of planar graphs and the class of forests is the class of graphs embeddable in the plane in such a way that all vertices lie on the exterior face. These have been called <u>outerplanar</u> graphs by Chartrand and Harary [11], who noted that there are similarities in characterizations of the three classes by means of excluded subgraphs. Whereas planar graphs have no subgraph homeomorphic to K_5 or $K_{3,3}$, and whereas forests have no subgraph homeomorphic to K_3 (or $K_{2,2}$, equivalently), outerplanar graphs are characterized by having no subgraph homeomorphic to K_4 or $K_{2,3}$, except the graph obtained by deleting an edge from K_4.

No covering problems involving outerplanar graphs have been investigated to any extent, but exact results have been found for packing K_p and $K(m,n)$ with non-outerplanar graphs. Proofs of these results will be given here.

Both K_4 and $K(2,3)$ have precisely 6 edges, so an upper bound of $[\frac{q}{6}]$ is immediate for the number of non-outerplanar graphs in a packing of a graph with q edges. For K_p and $K(m,n)$, this will turn out to be the maximum number, and no proper homeomorphs of either K_4 or $K(2,3)$ will be required in a packing. Notation we will use is that $\langle u\ v\ w\ x \rangle$ denotes the complete graph on vertices u, v, w, and x, and $\langle {u\ v\ w \atop x\ y} \rangle$ denotes the complete bipartite graph $K(2,3)$ in which x and y are adjacent to u, v, and w.

<u>Theorem 12</u>. The maximum number of non-outerplanar graphs in a packing of $K(m,n)$, for $m > 1$ and $n > 1$, is $[\frac{mn}{6}]$.

<u>Proof</u>. The value of the upper bound has already been established. We will give a construction yielding $[\frac{mn}{6}]$ edge-disjoint $K(2,3)$ subgraphs in $K(m,n)$.

The cases in which $2 \leq m \leq n \leq 7$ will be considered first. There are twenty-one of these cases, but many can be handled in small groups.

The cases $K(2,2)$ and $K(2,3)$ are obvious. Furthermore, by partitioning the vertices of one set into pairs or triples, one clearly has an exact decomposition of $K(2,6)$, $K(3,4)$, and $K(3,6)$ into $K(2,3)$'s.

The graphs $K(2,4)$, $K(2,5)$, and $K(3,3)$ have fewer than six edges more than $K(2,3)$, so can contain no more $K(2,3)$'s. The same applies to $K(3,5)$ and $K(4,4)$ with respect to $K(3,4)$.

The cases $K(4,5)$ and $K(5,5)$ require special constructions to show packings of 3 and 4 $K(2,3)$'s respectively. Here the vertices in one set are denoted by numbers, the other by letters.

$K(4,5)$: $\quad \langle {1\ 2 \atop A\ B\ C} \rangle \qquad\qquad \langle {3\ 4 \atop A\ B\ C} \rangle \qquad\qquad \langle {1\ 2\ 3 \atop D\ E} \rangle$

$K(5,5)$: $\quad \langle {1\ 2\ 3 \atop A\ B} \rangle \qquad \langle {3\ 4\ 5 \atop D\ E} \rangle \qquad \langle {1\ 2 \atop C\ D\ E} \rangle \qquad \langle {4\ 5 \atop A\ B\ C} \rangle$

Choosing $a \geq 0$ and $b \geq 0$ so that $r = 2a + 3b$, we note that $K(r,6)$ is the union of $2a$ $K(2,3)$'s in which the three vertices come from the original set of 6, and of $3b$ $K(2,3)$'s in which the two come from the set of 6. We note that this construction clearly applies for all $r > 1$.

Each case $K(r,7)$ for $2 \leq r \leq 5$ follows from the exact decomposition of $K(r,6)$ since the difference in edges is fewer than six.

The remaining case is $K(7,7)$, which contains 8 $K(2,3)$'s in this way

$$K(7,7) : \quad \langle{}^{1}_{A}{}^{2}_{B}{}_{C}\rangle \quad \langle{}^{1}_{D}{}^{3}_{E}{}_{F}\rangle \quad \langle{}^{4}_{A}{}^{5}_{B}{}_{F}\rangle \quad \langle{}^{6}_{C}{}^{7}_{D}{}_{E}\rangle$$

$$\langle{}^{3}_{C}{}^{4}_{}{}^{5}_{G}\rangle \quad \langle{}^{2}_{F}{}^{6}_{}{}^{7}_{G}\rangle \quad \langle{}^{2}_{D}{}^{4}_{}{}^{5}_{E}\rangle \quad \langle{}^{3}_{A}{}^{6}_{}{}^{7}_{B}\rangle$$

The result for the general case $K(m,n)$ now follows by letting $m = 6h + s$ with $2 \le s \le 7$ and $n = 6k + t$ with $2 \le t \le 7$. The edges of $K(m,n)$ can then be partitioned into subgraphs $K(6h,6k)$, $K(6h,t)$, $K(s,6k)$, and $K(s,t)$. By the earlier observation regarding $K(r,6)$, each of the first three of these is completely decomposable into $K(2,3)$'s. Since $K(s,t)$ contains $[\frac{st}{6}]$ $K(2,3)$'s, the proof is complete.

<u>Theorem 13</u>. The maximum number of non-outerplanar graphs in a packing of K_p is $[\frac{p(p-1)}{12}]$.

<u>Proof</u>. Since K_p has $p(p-1)/2$ edges, the upper bound is immediate. Its exactness for value of p from 2 to 13 will be shown first. When $p \le 5$, the verification is trivial. For $6 \le p \le 13$, the following table gives a construction; the number following K_p is both the value of $[\frac{1}{6}\binom{p}{2}]$ and the number of graphs in the corresponding packing.

K_6: 2 $\langle 1\ 2\ 3\ 4\rangle$ $\langle{}^{5}_{1}{}^{6}_{2}{}_{3}\rangle$

K_7: 3 $\langle 1\ 2\ 3\ 4\rangle$ $\langle{}^{1}_{5}{}^{2}_{6}{}_{7}\rangle$ $\langle{}^{3}_{5}{}^{4}_{6}{}_{7}\rangle$

K_8: 4 $\langle 1\ 2\ 3\ 4\rangle$ $\langle 5\ 6\ 7\ 8\rangle$ $\langle{}^{1}_{5}{}^{2}_{6}{}_{7}\rangle$ $\langle{}^{3}_{5}{}^{4}_{6}{}_{7}\rangle$

K_9: 6 $\langle 1\ 2\ 3\ 4\rangle$ $\langle 4\ 5\ 6\ 7\rangle$ $\langle 7\ 8\ 9\ 1\rangle$ $\langle{}^{2}_{5}{}^{3}_{6}{}_{7}\rangle$ $\langle{}^{5}_{8}{}^{6}_{9}{}_{1}\rangle$ $\langle{}^{8}_{2}{}^{9}_{3}{}_{4}\rangle$

K_{10}: 7 $\langle 1\ 2\ 3\ 4\rangle$ $\langle 5\ 6\ 7\ 8\rangle$ $\langle{}^{5}_{4}{}^{6}_{9}{}_{10}\rangle$ $\langle{}^{7}_{4}{}^{8}_{9}{}_{10}\rangle$ $\langle{}^{5}_{1}{}^{6}_{2}{}_{3}\rangle$ $\langle{}^{7}_{1}{}^{8}_{2}{}_{3}\rangle$ $\langle{}^{9}_{1}{}^{10}_{2}{}_{3}\rangle$

K_{11}: 9 K_9 construction and $\langle{}^{10}_{1}{}^{11}_{2}{}_{3}\rangle$ $\langle{}^{10}_{4}{}^{11}_{5}{}_{6}\rangle$ $\langle{}^{10}_{7}{}^{11}_{8}{}_{9}\rangle$

K_{12}: 11 K_9 construction and $\langle 9\ 10\ 11\ 12\rangle$ $\langle{}^{1}_{10}{}^{2}_{11}{}_{12}\rangle$ $\langle{}^{3}_{10}{}^{4}_{11}{}_{12}\rangle$

$\langle{}^{5}_{10}{}^{6}_{11}{}_{12}\rangle$ $\langle{}^{7}_{10}{}^{8}_{11}{}_{12}\rangle$

K_{13}: 13 K_9 construction and

$\langle 10\ 11\ 12\ 13\rangle$ $\langle{}^{10}_{1}{}^{11}_{2}{}_{3}\rangle$ $\langle{}^{10}_{4}{}^{11}_{5}{}_{6}\rangle$ $\langle{}^{10}_{7}{}^{11}_{8}{}_{9}\rangle$ $\langle{}^{12}_{1}{}^{13}_{2}{}_{3}\rangle$ $\langle{}^{12}_{4}{}^{13}_{5}{}_{6}\rangle$

$\langle{}^{12}_{7}{}^{13}_{8}{}_{9}\rangle$

For any value of p > 13, let n and r be such that p = 12n + r with 2 ≤ r ≤ 13. Partition the vertices into n subsets of 12 vertices each and one subset of r vertices. From Theorem 12, the result on complete bipartite graphs, it follows that the edges joining vertices of different sets can be completely partitioned into K(2,3)'s. Into each complete graph on 12 vertices, 11 graphs of the form K_4 or K(2,3) can be packed as above with no edges wasted, and into K_r, $[\frac{r(r-1)}{12}]$ can be packed. Thus, the edges of K_p can be divided into K(2,3)'s and K_4's with fewer than 6 edges left over, which completes the proof of the theorem.

9. <u>Packing with complete bipartite graphs</u>. One result obtained in the proof of Theorem 11 gives a maximum packing of K(m,n) with K(3,3)'s as a lower bound for the coarseness. Also, a packing of K(m,n) with K(2,2)'s turns out to yield the optimal packing with cycles, given in Theorem 9. These are both special cases of the problem of packing K(m,n) with a regular complete bipartite graph K(r,r), which has been solved completely [6].

<u>Theorem 14</u>. The maximum number of K(r,r)'s in a packing of K(m,n) is min $([\frac{m}{r} [\frac{n}{r}]], [[\frac{m}{r}] \frac{n}{r}])$.

<u>Conclusion</u>. This survey was not intended to be an exhaustive compilation of covering and packing problems involving graphs. We have concentrated on those problems for which the solutions are known for most of the complete and complete bipartite graphs. There are some problems involving just these which are interesting and remain unsolved in general, as for example, packing K_p with K_r or with K(n,n). Some solutions are known when p and r satisfy certain conditions, but the only general solution is for packing K_p with triangles. Theorem 8 is a corollary of this and the solutions are the same except when p ≡ 5 (mod 6) in which case the number of triangles in a packing is one fewer than the number of cycles. Guy [14] considers other problems of this type.

REFERENCES

1. J. Battle, F. Harary, and Y. Kodama, Every planar graph with nine points has a non-planar complement. <u>Bull. Amer. Math. Soc.</u> 68 (1962), 569-571.

2. L.W. Beineke, Decompositions of complete graphs into forests. <u>Publ. Math. Inst. Hungar. Acad. Sci.</u> 9 (1964), 589-594.

3. L.W. Beineke, Complete bipartite graphs: Decomposition into planar subgraphs. Chapter in <u>A Seminar on Graph Theory</u>, F. Harary, Ed. Holt, Rinehart and Winston, New York, 1967, pp. 42-53.

4. L.W. Beineke, The decomposition of complete graphs into planar subgraphs. Chapter in <u>Graph Theory and Theoretical Physics</u>, F. Harary, Ed. Academic Press, London, 1967, pp. 139-153.

5. L.W. Beineke, Minimal decompositions of complete graphs into sub-
    graphs with embeddability properties. <u>Canad. J. Math</u>. 21 (1969).
    (to appear)

6. L.W. Beineke, Packings of complete bipartite graphs. (to appear)

7. L.W. Beineke and R.K. Guy, The coarseness of the complete bipar-
    tite graph. <u>Canad. J. Math</u>. 21 (1969) (to appear).

8. L.W. Beineke and F. Harary, The thickness of the complete graph.
    <u>Canad. J. Math</u>. 17 (1965), 850-859.

9. L.W. Beineke, F. Harary, and J.W. Moon, On the thickness of the
    complete bipartite graph. <u>Proc. Camb. Phil. Soc</u>. 60 (1964), 1-5.

10. G. Chartrand, D. Geller, and S. Hedetniemi, Graphs with forbidden
    subgraphs. <u>J. Combinatorial Theory</u> (to appear).

11. G. Chartrand and F. Harary, Planar permutation graphs. <u>Ann. Inst.</u>
    <u>H. Poincare</u> Sect B (N.S.) 3 (1967), 433-438.

12. P. Franklin, A six-color problem, <u>J. Math. Phys</u>. 13 (1934), 363-
    369.

13. R.K. Guy, A coarseness conjecture of Erdos. <u>J. Combinatorial</u>
    <u>Theory</u> 3 (1967), 38-42.

14. R.K. Guy, A many-facetted problem of Zarankiewicz. (this volume)

15. R.K. Guy and L.W. Beineke, The coarseness of the complete graph.
    <u>Canad. J. Math</u>. 20 (1968), 888-894.

16. A.M. Hobbs and J.W. Grossman, Thickness and connectivity in
    graphs. (to appear)

17. J. Mayer, L'epaisseur des graphes complets K_{34} et K_{40}. J. <u>Com-</u>
    <u>binatorial Theory</u>. (to appear)

18. C. St. J. A. Nash-Williams, Edge-disjoint spanning trees of fin-
    ite graphs. <u>J. London Math. Soc</u>. 36 (1961), 445-450.

19. C. St. J. A. Nash-Williams, Decomposition of finite graphs into
    forests. <u>J. London Math. Soc</u>. 39 (1964), 12.

20. G. Ringel, Die toroidale Dicke des vollständigen Graphen. <u>Math.</u>
    <u>Zeitschr</u>. 87 (1965), 19-26.

21. W.T. Tutte, On the problem of decomposing a graph into n con-
    nected factors. <u>J. London Math. Soc</u>. 142 (1961), 221-230.

22. H. Walther, Über die Zerlegung des vollständigen Graphen in paare
    planare Graphen. Chapter in <u>Beiträge zur Graphentheorie</u>, H. Sachs,
    H.-J. Voss, and H. Walther, Eds. B.G. Teubner Verlag, Leipzig,
    1968. pp. 189-205.

SECTION GRAPHS FOR FINITE PERMUTATION GROUPS

I.Z. Bouwer, University of Waterloo

Our aim here is to show that by imbedding a graph in a l[']ger graph the symmetry on it may be curtailed to any desired exten... The results appear in [2] but are presented here in a more expository fashion.

1. Introduction. All the structures considered will be finite. An automorphism of a graph is a permutation of its vertices which preserves adjacency. The automorphisms of a graph X form a permutation group, called the (symmetry) group of X, which we denote by A(X).

The structure of the group of a graph may be viewed on two levels: its structure as an abstract group, and its structure as a permutation group. The question thus arises as to how representative the groups of graphs are of these structures. In particular, we may ask:

(i) Given any group G: does there exist a graph X such that A(X) is (abstractly) isomorphic to G?

(ii) Given any permutation group P: does there exist a graph X such that A(X) is isomorphic (as a permutation group) to P?

We recall that two permutation groups P and Q, with respective object sets W and V, are isomorphic as permutation groups, if there exists a one-to-one mapping f from W onto V, and a group isomorphi m $\alpha: P \to Q$, such that

$$\alpha(p) \circ f = f \circ p \tag{1}$$

for all $p \in P$. From (1) we note that if the elements of V are allowed to assume the labels of their preimages under f, that is, if f is postulated as the identity, then α is the identity isomorphism. Thus, permutation groups which are isomorphic, as permutation groups, can differ only in the respect that their objects are labeled differently.

Question (i) was put in 1936 by König [12, p. 5] and solved by Frucht [5], who showed that for any given group G the Cayley colour graph of G may be modified so as to yield (an infinity of nonisomorphic) graphs X with the property that A(X) is (abstractly) isomorphic to G. Thus the (abstract) groups of graphs represent all groups. In fact, they represent them in abundance, in the sense that various properties for X may be prescribed (for instance: connectivity, chromatic number (≥ 2), and degree of regularity (≥ 3)), and still possess an infinity of solutions [10][16].

For permutation groups the situation is different [5, p. 247]. The condition that a given permutation group P is isomorphic, as a permutation group, to the group of a graph X, entails not only that X has as many vertices as there are elements in the object set of P, but that the edges of X so interconnect the vertices that only the permutations in P occur as the automorphisms of X. It is therefore not surprising to find that counterexamples are plentiful [9][11].

For the sake of interest we describe one set of counterexamples to (ii). A permutation group is called 2 <u>set-transitive</u> [1] if for each pair of two-element subsets S, T of its object set, there is a permutation in the group which transforms S to T. In particular, all k-transitive (k ≥ 2) permutation groups are 2 set-transitive. Let Q denote any 2 set-transitive proper subgroup (for instance, the alternating group A_n) of the symmetric group S_n (n ≥ 3). If a graph X is such that the elements of Q occur as automorphisms of X, then the 2 set-transitivity of Q implies that if X has at least one edge, then each two vertices of X are joined by an edge, so that X is the complete graph on n vertices. Otherwise X is the null graph (i.e., the graph with no edges) on n vertices. In both cases: $A(X) = S_n$. Thus, at best, Q occurs as a proper subgroup of the group of a graph.

The problem of characterizing the permutation groups which are isomorphic, as permutation groups, to groups of graphs, seems to be difficult, and only some partial results have been obtained [3][4][9] [11][13][14][17]. Since to any given permutation group P there corresponds at most a finite number m(P), say, of graphs X such that A(X) is isomorphic, as a permutation group, to P, the problem arises [7, p. 196] to evaluate m(P). Progress has been made by Sheehan [18], who describes an algorithm (in terms of permutation representations) which may be applied to any given P to yield m(P).

Here we shall not consider these problems, but instead formulate (and will answer in the affirmative) the following weaker form of (ii): (iii) Given any permutation group P: does there exist a graph X such that P occurs as a subdirect constituent of A(X)?

We illustrate with the permutation group:

R_4 = (1234)cyc = {1=(1)(2)(3)(4), (1234), (13)(24), (1432)}.

R_4 is a member of the family of permutation groups (12 ... n)cyc (n ≥ 3), which are known to have no graphs [11, p. 509, Corollary 1]. Indeed, up to complementation, the only graphs Y such that R_4 ⊆ A(Y), are those given by Figures 1 and 2, whose respective groups are the dihedral and symmetric groups, both of which properly contain R_4.

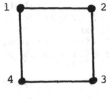

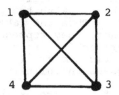

Figure 1   Figure 2

Now let Y denote any one of these graphs, say the graph of Figure 2. In Figure 3 the graph Y is shown extended to a supergraph X, which contains Y as a section graph, that is, only the edges of Y join vertices of Y (see [15]).

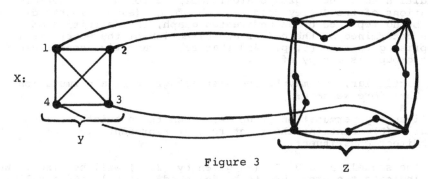

X:

y

Figure 3

Z

The section graph Z of X, defined on the vertices not belonging to Y, is a graph whose automorphisms can be described, in terms of its drawing in Figure 3, as the rotations through multiples of 90°. (In the literature Z occurs [8] as the smallest graph whose group is abstractly isomorphic to the cyclic group of order 4.) Since in X the vertices belonging to Y are the only ones of degree 4, we see that they are mutually permuted by any automorphism of X, that is, the restriction A(X)|V(Y) of A(X) to the vertex set V(Y) of Y is defined. This restriction must of course be a permutation sub-group of A(Y). Since the symmetry of Z only allows the "rotations through multiples of 90°", and since, with respect to automorphisms of X, the edges interconnecting Y and Z transfer these rotations to Y, we have that the restriction A(X)|V(Y) is the permutation group R_4. We may say that the symmetry of Y in excess of R_4 has been absorbed in the larger structure X, and that only that given by R_4 has been retained on Y.

The result we present, states that this can be done in general:

Theorem. Given any couple of the form (P, Y), where P is a per-mutation group, and Y is a graph such that P ⊆ A(Y). Then there exists a supergraph X of Y such that:
  (a) Y is a section graph of X,
  (b) The restriction A(X)|V(Y) of A(X) to the vertex set
V(Y) of Y is defined, and is isomorphic, as a permutation group, to P.

Condition (b) may also be expressed by the statement that V(Y) is invariant under A(X), and that the subdirect constituent of A(X) on V(Y) is a copy of P. Since for any permutation group P one can find a graph Y (for instance, the complete graph) such that P ⊆ A(Y), the theorem provides an affirmative answer to (iii).

2. Proof of Theorem. In the right regular representation, G → G* of a group G each element g of G is represented by the permuta-tion g*: h → hg (h ∈ G) of the elements h of G. The Cayley col-our graph of G (i.e., the edge-coloured directed graph whose ver-tices are the elements of G, to each ordered pair (a, b) of which there corresponds an edge directed from a to b, and assigned the "colour" ba^{-1}) has the property that its group of (colour-preserv-

ing) automorphisms is the right regular representation G^* of G. Since the graphs which Frucht constructed in [5], and which satisfy the condition that their groups are isomorphic to G, are certain modifications of the Cayley colour graph of G (each coloured directed edge being replaced by an undirected graph, the definition of which is determined by the colour and direction of the edge), it is not surprising to find [5, p. 245] that each transitive constituent of their groups is a copy of G^*.

In particular, G may be the abstract group of a permutation group. Therefore we have:

(2.1) Given any permutation group P. Then there exists a graph F such that $A(F)$ contains the right regular representation P^* of P as a subdirect constituent.

In the sequel P and F (as given by (2.1)) will be fixed. We briefly indicate the procedure to be followed. Step 1: An auxiliary graph Z is constructed whose vertex set can be partitioned into two sets, denoted by W and R, such that $A(Z)$ contains the subdirect product,

$$\{(p, p^*) : p \in P\}$$

of a copy of P acting on W, and a copy of P^* acting on R. Step 2: The graph Z is attached to F by identifying R with any one, say D, of the orbits of $A(F)$ on which a copy of P^* acts (the identification being such that the copies of P^* on D and on R coincide). Step 3: A supergraph X of the combined graph is constructed such that its section graph on W is the graph Y (see Theorem), and such that $A(X)|W$ is defined, and isomorphic, as a permutation group, to P.

Step 1. Let W be the object set of P, with $|W| = n$. Let $W^{(n)}$ denote the set of all ordered sequences $(x_1, \ldots, x_n)$ (no repetitions) of the n elements of W. Each $p \in P$ induces a unique permutation

$$p^{(n)} : (x_1, \ldots, x_n) \to (p(x_1), \ldots, p(x_n))$$

on $W^{(n)}$, and the mapping, $p \to p^{(n)}$ defines an (abstract) isomorphism of P onto the permutation group $P^{(n)} = \{p^{(n)} : p \in P\}$. We choose R to be any fixed orbit of $P^{(n)}$, and note here that the restriction $P^{(n)}|R$ is isomorphic, as a permutation group, to P^* (see (2), which may serve as a mapping f in an application of (1)).

Let the orbits of P be $W_1, W_2, \ldots, W_s$.

The graph Z is constructed as follows; its vertex set is the (disjoint) union of W and R, and its edges are formed by joining $x \in W_i$ and $a \in R$ if and only if x occurs as the first of those entries of the n-tuple a that belong to W_i $(i = 1, \ldots, s)$. In particular, the section graphs of Z having W and R as their vertex sets, are null graphs.

As an immediate consequence of the definition of Z we have:

(2.2) A(Z) contains the subdirect product

$$\{(p, p^{(n)}|R) : p \in P\}$$

of a copy of P acting on W, and a copy of P* acting on R.

From the fact that in Z each vertex in R is joined to exactly one vertex in each W_i (i = 1,...,s), while each vertex in W is joined to at least one vertex in R, it follows that an automorphism of Z which leaves invariant each of the sets W_1, ..., W_s, is uniquely determined by its restriction on the vertices in R. Thus (2.2) implies:

(2.3) If an automorphism σ of Z leaves invariant each of the sets W_1, ..., W_s, and if $\sigma|R$ has the form $p^{(n)}|R$ for some $p \in P$, then $\sigma|W = p$.

Step 2. The graph Z is attached to F as follows: Let D be any orbit of A(F) on which a copy of P* acts. The mapping from D to R given by:

$$p^*(d) \rightarrow p^{(n)}(r), \quad p \in P, \tag{2}$$

where d and r are any fixed elements of D and R, respectively, is one-to-one and onto. The sets D and R may therefore be identified with respect to (2) (in which case, for any $p \in P$, the permutation p* on D identifies with $p^{(n)}|R$). The two graphs F and Z then combine to form a graph L whose vertex set is the union of W and the vertex set of F, and whose edge set is the union of the edge sets of Z and F. In particular, the section graph of L with vertex set W is the null graph.

Since each automorphism σ of F, when restricted to D (identified with R) is an element of the form p* (identified with $p^{(n)}|R$), where $p \in P$, it follows from (2.3) that if σ leaves invariant each of the sets W_1, ..., W_s, then its restriction to W is p. Also, from (2.2) and the definition of L, it follows that any permutation on W of the form $p \in P$ can be extended to an automorphism of L. Thus:

(2.4) Let B(L) be the permutation subgroup of A(L) consisting of all those automorphisms of L which leave invariant each of the sets W_1, ..., W_s. Then $B(L)|W = P$.

Step 3. We may write, V(Y) = W, where Y is the graph (given in the Theorem) satisfying, $P \subseteq A(Y)$. A supergraph J of L is now formed by inserting the edges of Y in the (null) section graph of L on W (leaving the rest of L unchanged). Since $P \subseteq A(Y)$, we have that the section graph Y does not obstruct the action of P on W, so that the equivalent of (2.4) holds true for J;

(2.5) Let B(J) be the permutation subgroup of A(J) consisting of all those automorphisms of J which leave invariant each of the sets W_1, ..., W_s. Then $B(J)|W = P$.

The graph J is close to being the graph X as required by the

Theorem. It satisfies condition (a) of the Theorem, and provided we can assure that the sets $W_1, \ldots, W_s$ are invariant under all automorphisms of the graph, condition (b) of the Theorem will be met (by (2.5)). Now there are many ways in which a supergraph X of J can be constructed, such that both Conditions (a) and (b) of the Theorem are met. One especially simple construction (modifying one given in [5, p. 242]) consists in adjoining to each vertex of each set W_i (i = 1, ..., s) a number m(i) of "tails of length 1" (see [5, p. 242]) such that the degree in X of any vertex in any given set W_j is different from the degree in X of each vertex of X not belonging to W_j. The sets $W_1, \ldots, W_s$ are accordingly distinguished by the degrees of their vertices, and it follows that they are invariant under all automorphisms of X. It is now an easy matter to verify that X has all the properties as required by the Theorem.

3. <u>Variations of Theorem</u>. For graphs which are not necessarily undirected, and which may have loops and multiple edges, an automorphism can be defined as a permutation of the vertices which preserves the section graph structure on ordered pairs of vertices. From our constructions it is clear that the Theorem holds in the case when the graphs X and Y are not in any way restricted. In [2] it is shown that it also holds when Y and X are restricted to be directed (with or without the condition that there are no loops or multiple edges). It is also shown there that in all cases there exist an infinity of (nonisomorphic) solutions X, where X may be assumed to have the additional properties that it is connected and that A(X) is abstractly isomorphic to P.

REFERENCES

1. R.A. Beaumont and R.P. Peterson, Set-transitive permutation groups, <u>Canad. J. Math</u>, 7 (1955), 34-42.

2. I.Z. Bouwer, Section graphs for finite permutation groups, <u>J. Combinatorial Theory</u>, to appear.

3. C.-Y. Chao, On a theorem of Sabidussi, <u>Proc. Amer. Math. Soc.</u>, 15 (1964), 291-292. (see also [6][13])

4. C.-Y. Chao, On groups and graphs, <u>Trans. Amer. Math. Soc.</u>, 118 (1965), 488-497.

5. R. Frucht, Herstellung von Graphen mit vorgegebener abstrakter Gruppe, <u>Compositio Math.</u>, 6 (1938), 239-250.

6. H.A. Gindler, Review 91, <u>Mathematical Reviews</u> 31 (1966).

7. F. Harary, Combinatorial problems on graphical enumeration, Chapter 6 in E.F. Bechenbach (Ed.): <u>Applied Combinatorial Mathematics</u>, Wiley, New York, 1964.

8. F. Harary and E.M. Palmer, The smallest graph whose group is cyclic, <u>Czechoslovak Math. J.</u>, 16 (91) (1966), 70-71.

9. R.L. Hemminger, On the group of a directed graph, _Canad. J. Math._, 18 (1966), 211-220.

10. H. Izbicki, Reguläre Graphen beliebigen Grades mit vorgegebenen Eigenschaften, _Monatsh. Math._, 64 (1960), 15-21.

11. I.N. Kagno, Linear graphs of degree ≤ 6 and their groups, _Amer. J. Math._, 68 (1946), 505-520. Corrections, _Amer. J. Math._, 77 (1955), 392.

12. D. König, _Theorie der endlichen und unendlichen Graphen_, Leipzig, 1936; reprinted by Chelsea, New York, 1950.

13. M.H. McAndrew, On graphs with transitive automorphism groups, _Notices Amer. Math. Soc._, 12 (1965), 575.

14. L.A. Nowitz, On the non-existence of graphs with transitive generalized dicyclic groups, _J. Combinatorial Theory_, 4 (1968), 49-51.

15. O. Ore, _Theory of graphs_, Amer. Math. Soc. Colloq. Publ. 38, Providence, 1962.

16. G. Sabidussi, Graphs with given group and given graph theoretical properties, _Canad. J. Math._, 9 (1957), 515-525.

17. G. Sabidussi, Vertex-transitive graphs, _Monatsh. Math._, 68 (1964), 426-438. (see also [6][13])

18. J. Sheehan, The number of graphs with a given automorphism group, _Canad. J. Math._, 20 (1968), 1068-1076.

Added in proof. Professor R. W. Frucht pointed out to the author that the graph Z in Figure 3 cannot be viewed (as claimed in [8]) as the "smallest" graph whose is abstractly isomorphic to the cyclic group C_4 of order 4. From a result of R. C. Meriwether, quoted in G. Sabidussi, Review #2563, Mathematical Reviews 33 (1967), it follows that the smallest number of vertices of a graph with group isomorphic to C_4, is 10 (and not 12), while Professor Frucht has found an example with only 18 (and not 20) edges.

NEARLY REGULAR POLYHEDRA WITH TWO
EXCEPTIONAL FACES

D.W. Crowe, University of Wisconsin

1. **Introduction**. In this paper we are concerned with 3-valent (cubic) convex polyhedra which are <u>nearly regular</u>, in the sense that all faces, with the exception of exactly two faces <u>a</u> and <u>b</u>, have edge numbers divisible by some fixed k (k = 3,4,5). The edge numbers of <u>a</u> and <u>b</u> will be denoted by $|a|$ and $|b|$.

Grünbaum [1, p. 272] has proved that a convex 3-valent polyhedron whose other faces all have edge numbers divisible by k (k = 3,4,5) cannot have exactly one exceptional face, and that if it has exactly two exceptional faces they cannot have a common edge. These facts will be referred to as Grünbaum's theorem. Malkevitch [2] has proved that the same is true for 4-valent and 5-valent polyhedra.

In the present paper we extend Grünbaum's theorem to the case where the two exceptional faces are joined by a simple path of length n (≤ 4), by determining congruence relations (mod k) which must hold between $|a|$ and $|b|$ if there is such a path joining them. The symmetry of the results obtained suggests strongly that our theorem is the first step in an induction proof of suitable theorems <u>for all n</u>. In Section 3 we give such an inductive proof for an especially simple case. Further results of this type will be given in a later paper.

Implicit throughout in the following is Steinitz's theorem [1, p. 235] to the effect that a graph G having at least four vertices is the network of vertices and edges of a convex polyhedron if and only if G is planar and 3-connected. Thus our proofs can be carried out by looking at planar graphs rather than the polyhedra themselves.

2. **Main Results**. We consider planar, 3-valent and 3-connected graphs G, in which the edge number of every face is divisible by k (k = 3,4,5), except for exactly two exceptional faces <u>a</u>, <u>b</u>. If there is a simple path of n edges, whose endpoints (and no other points) are vertices of <u>a</u> and <u>b</u> respectively we denote G by G(k, 2, n). Thus, for example, Grünbaum's theorem [1, p. 272] implies that G(k, 2, 0) does not exist. Note that a given graph may well be both a G(k, 2, n_1) and a G(k, 2, n_2) for $n_1 \neq n_2$.

<u>Theorem 1</u>. (0) G(k, 2, 0) does not exist (Grünbaum).
(1) G(k, 2, 1) does not exist unless k = 4 and $|a| \equiv |b|$ (mod 4).
(2) G(k, 2, 2) does not exist unless k = 3 and $|a| + |b| \equiv 0$ (mod 3).
(3) G(k, 2, 3) does not exist unless
(i) k = 4 and $|a| + |b| \equiv 0$ (mod 4) or
(ii) k = 5 and $|a| \equiv |b|$ (mod 5).
(4) G(k, 2, 4) does not exist unless k = 4 or 5 and $|a| + |b| \equiv 0$ (mod k).

Furthermore, for n $\leq$ 4, graphs G(k, 2, n) can be constructed of each type not specifically excluded by the theorem.

Proof of (1). Suppose, if possible, k = 3. Then the two exceptional faces must be arranged as in the first row of Figure 1, where faces
are labelled by their number of edges, reduced mod 3. By "shaving"
the joining edge once or twice

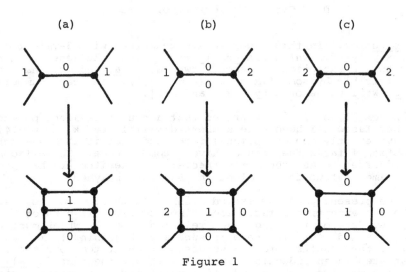

Figure 1

the configurations of the second row of Figure 1 are obtained. Cases
(a) and (b) contradict (0), and case (c) is impossible by Grünbaum's
theorem since it has a single exceptional face.

In case k = 4, a t-gonal prism is a G(4, 2, 1) with $|a|$ =
$|b|$ = t. We have only to show that the three other cases, $|a| \not\equiv |b|$
(mod 4), are impossible.

The reductions are shown in Figure 2.

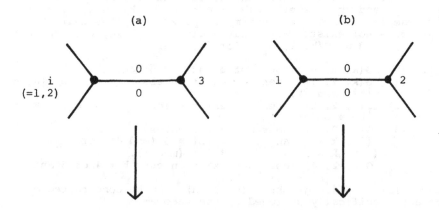

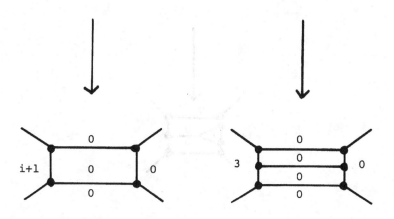

Figure 2

The case $k = 5$ is impossible. There are 10 cases to consider, and each reduces to an impossibility, as shown in Figure 3.

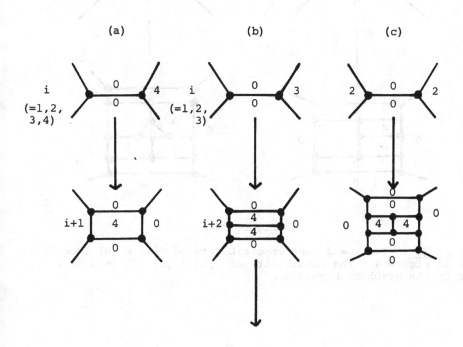

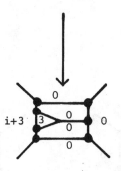

(d)   (e)

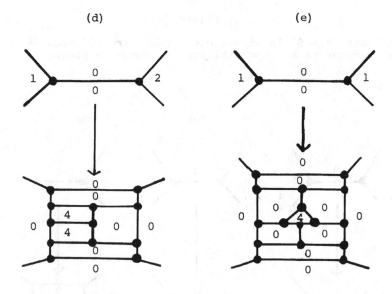

Figure 3

<u>Proof of (2)</u>. If k = 3 the impossibility of |a| ≡ |b| (mod 3) is
shown in Figure 4. The possibility of |a| + |b| ≡ 0 (mod 3) is
shown by the graph of Figure 5.

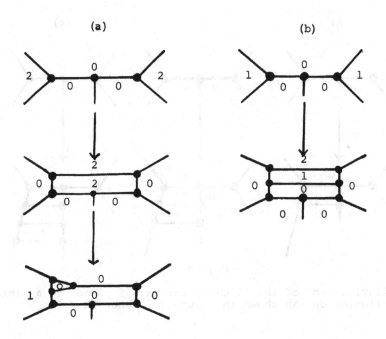

Figure 4

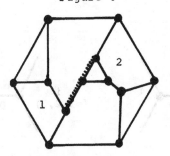

Figure 5

Each of the six cases for k = 4 leads to an impossible configuration by "shaving" one of the connecting edges one, two or three
times, as shown in Figures 6 (a), (b), (c). Hence k ≠ 4.

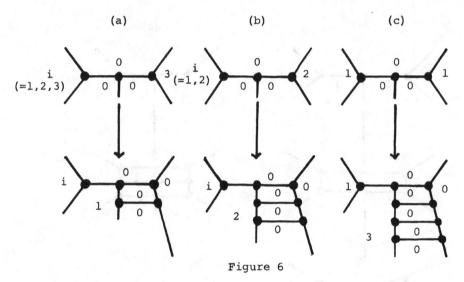

Figure 6

Similarly, each of the 10 cases for k = 5 leads to an impossible configuration, as shown in Figure 7. Hence k ≠ 5.

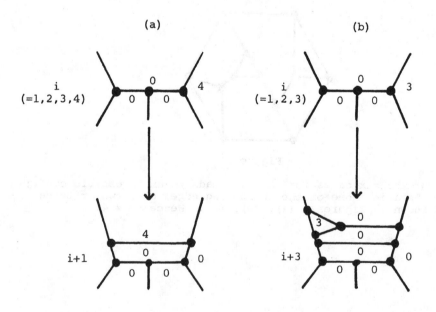

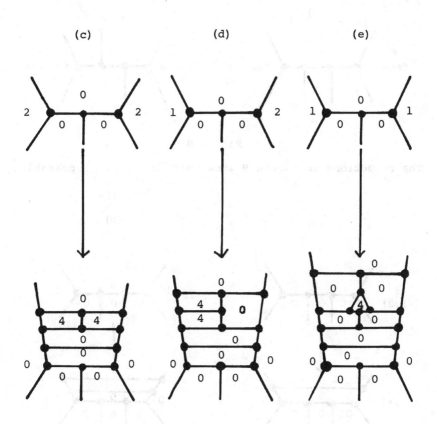

Figure 7

This completes the proof of (2).

Proof of (3). The number of cases doubles here, since there are two ways in which the exceptional faces a and b can be joined. (See

Figure 8).

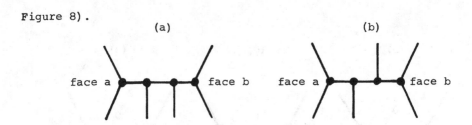

(a)   (b)

face a   face b   face a   face b

Figure 8

The reductions in Figure 9 show that k = 3 is impossible.

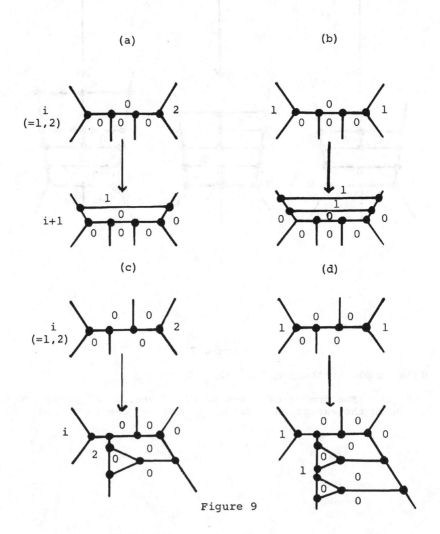

Figure 9

When k = 4 the six cases of the form shown in Figure 10 are all impossible. For, if |b| ≡ j (mod 4) then shaving edge e 4-j times leads to the impossible configuration of exactly two exceptional faces (a and c) having a common edge.

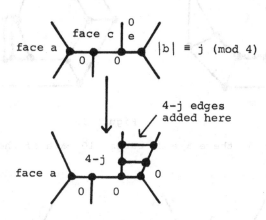

Figure 10

Of the other six cases for k = 4, the reductions (to case (1)) in Figure 11 show that four are impossible. That the remaining two (|a| + |b| ≡ 0 (mod 4)) are possible is shown by the graphs in Figure 12.

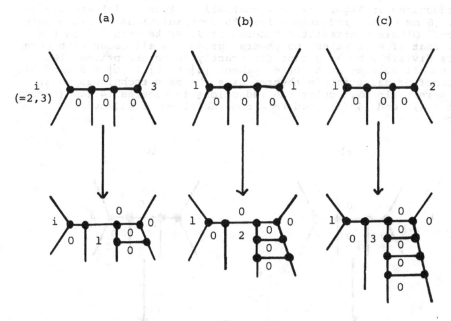

Figure 11

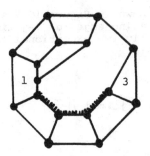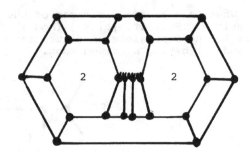

Figure 12

For k = 5 there are 20 cases, 10 each of the two types shown in Figure 13.

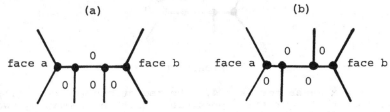

Figure 13

The reductions in Figure 14 show that all of type 13 (a) are impossible. Some of the reductions lead to contradictions of Grünbaum's theorem. Others contradict a theorem of J. Malkevitch [2] to the effect that if a trivalent polyhedral graph has all faces with edge numbers divisible by 5, except for exactly three exceptional faces a, b, c all meeting at a vertex, then $|a| \equiv |b| \equiv |c| \equiv 2$ (mod 5). (The proof of Malkevitch's theorem uses the same techniques we are using here). In particular the reductions in Figures 14 (a) (except for i = 4) and 14 (f) lead to a contradiction of Malkevitch's theorem.

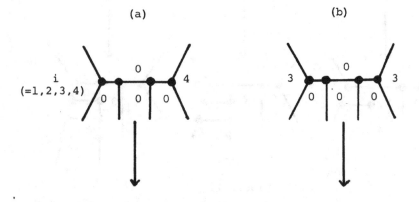

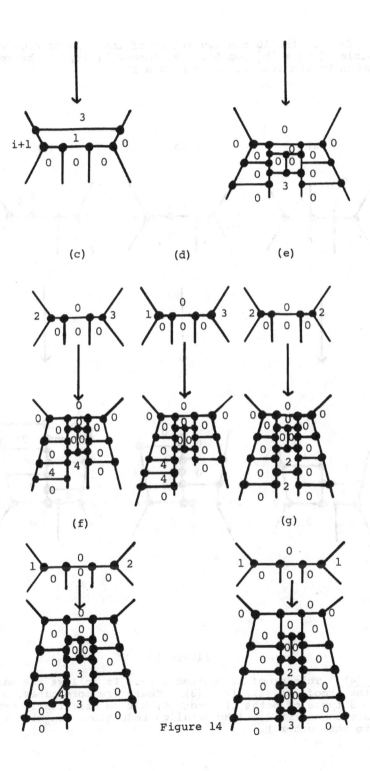

Figure 14

Only four of the 10 configurations of the type of Figure 13 (b) are possible ($|a| \equiv |b| \mod 5$). The impossibility of the remaining six is shown by the reductions in Figure 15.

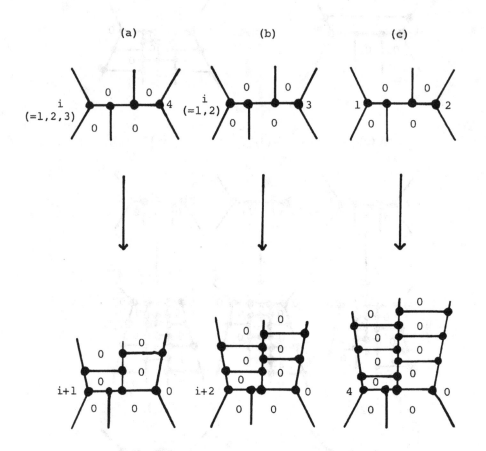

Figure 15

Proof of (4). This proof is omitted here. It follows the same pattern as the proofs of (1), (2), (3). There are more cases, since there are 3 types of paths of length 4, but no essentially new difficulties are encountered. (The detailed reductions can be obtained by writing to the author.)

3. <u>Geodesic paths of length</u> <u>n</u>. A path in a trivalent planar graph
is said to be <u>geodesic</u> if, of the two choices available to it upon en-
tering any given vertex, it alternately takes left and right. For example, the path (of length 3) joining face <u>a</u> to face <u>b</u> in Figure
13 (b) is a geodesic, while the path in Figure 13 (a) is not. This
terminology enables us to state

<u>Theorem 2</u>. If a $G(3, 2, n)$ contains a geodesic path of n edges
joining the two exceptional faces <u>a</u> and <u>b</u>, then $n \equiv 2 \pmod 4$ and
$|a| + |b| \equiv \pmod 3$.

<u>Proof</u>. Let $n = 4m + t$ $(t = 0,1,2,3)$ and proceed by induction on
m. The case $m = 0$ is proved by (0), (1), (2), (3) in Theorem 1.
The case $m + 1$ is reduced to the case m as in Figure 16. This
completes the proof.

(a)

(b)

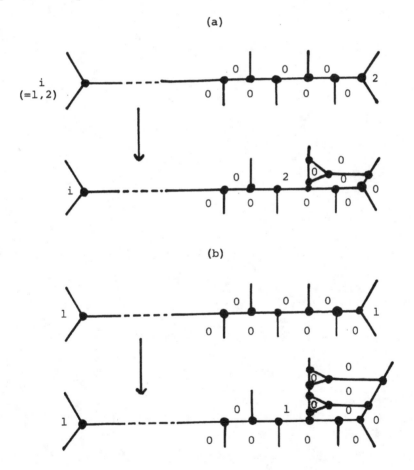

Figure 16

REFERENCES

1. B. Grünbaum, <u>Convex Polytopes</u>, Wiley, New York, 1967.

2. J. Malkevitch, Properties of planar graphs with uniform vertex
    and face structure, Ph.D. Thesis, University of Wisconsin, 1968.

SOME APPLICATIONS OF GRAPH THEORY TO NUMBER THEORY

Paul Erdős, Hungarian Academy of Science

The problems which we will discuss in this paper deal with sequences of integers; they are all of a combinatorial nature and graph theoretic results can be applied to some of them.

First we define the concept of an r-graph (for $r = 2$ we obtain the ordinary graphs). The elements of the r-graph are its vertices, some of whose r-tuples belong to our r-graph. $G_r(n, t)$ denotes an r-graph of n vertices and t r-tuples. $G_r(n; \binom{n}{r})$ denotes the complete r-graph $K_r(n)$ and $K_r(p_1, \ldots, p_r)$ denotes the r-graph of $(p_1 + \ldots + p_r)$ vertices with p_i vertices of the i-th class where each r-tuple all whose vertices are in different classes belongs to our $K_r(p_1, \ldots, p_r)$. Throughout this paper, "graph" will indicate a 2-graph. We will denote 2-graphs by G (i.e., in G_2 the index 2 will be omitted).

It is well known and easy to see that if $a_1 < \ldots < a_k < n$ and no a_i divides any other a_j then $\max k = [\frac{n + 1}{2}]$. Also if we assume that no a_i divides the product of all the other a_j's we can easily show that $\max k = \Pi(n)$. The same result holds if we assume that all the products $\prod_{i=1}^{k} a_i^{\alpha_i}$ are distinct (the α_i's being non-negative integers).

Let us now assume that our sequence has the property that no a_i divides the product of two other a_j's. I proved [3] that in this case

(1) $\quad \Pi(x) + c_1 x^{2/3}/(\log x)^2 < \max k < \Pi(x) + c_2 x^{2/3}/(\log x)^2.$

We outline the proof of the upper bound of (1). A simple lemma states that every integer $m \le x$ can be written in the form $u \cdot v$ where u is either a prime or is less than $x^{2/3}$ and v is less than $x^{2/3}$. Corresponding to the sequence $a_1 < \ldots < a_k$ we form a graph as follows: The vertices of our graph are the integers $< x^{2/3}$ and the primes p, $x^{2/3} < p \le x$. Put $a_i = u_i v_i$ by our lemma and let a_i correspond to the edge joining the vertices u_i and v_i. Our graph contains no path of length three (since no a_i divides the product of two other a_j's); thus our graph is a tree and thus has fewer edges than vertices or $k < \Pi(x) + x^{2/3}$. The inequality

$k < \Pi(x) + c_2 \, x^{2/3}/(\log x)^2$ can be obtained by an improvement of the lemma (not all the integers $< x^{2/3}$ are needed in the representation $m = uv$).

The lower bound in (1) uses Steiner triples. It would be interesting to sharpen (1) and prove that for a certain absolute constant c

(2) $$\max k = \Pi(x) + c \, x^{2/3}/(\log x)^2 + o(\frac{x^{2/3}}{(\log x)^2})$$

I have not been able to prove (2).

A generalization of the method which we used in the proof of (1) leads to the following more general result: Let $a_1 < \ldots < a_k \leq x$ be a sequence of integers where no a_i divides the product of r other a_j's. Then

(3) $$\Pi(x) + c_1^{(r)} \, x^{2/(r+1)}/(\log x)^2 < \max k < \Pi(x) + c_2^{(r)} \, x^{2/(r+1)}/(\log x)^2.$$

Assume now that our sequence $a_1 < \ldots < a_k \leq x$ is such that the products $a_i a_j$ are all distinct. Then [3] [4]

(4) $$\Pi(x) + c_4 \, x^{3/4}/(\log x)^{3/2} < \max k < \Pi(x) + c_3 \, x^{3/4}/(\log x)^{3/2}.$$

The proof of (4) again uses the lemma used in the proof of (1) and the graph theoretic representation of the sequence $a_1 < \ldots < a_k$. The fact that the products $a_i a_j$ are all distinct implies that the graph corresponding to the sequence $a_1 < \ldots < a_k$ contains no 4-cycle. The upper bound in (4) follows from the fact that every $G(n; [c_5 \, n^{3/2}])$ contains a rectangle. The lower bound is due to Miss E. Klein and myself and is easy to obtain using finite geometries [3].

Here I would like to mention a problem in graph theory which is not yet completely solved. Denote by $f(n)$ the smallest integer for which every $G(n; f(n))$ contains a 4-cycle. W. Brown and V.T. Sós, Rényi and I ([2], [5]) proved that

(5) $$f(n) = (1/2 + o(1))n^{3/2}.$$

We are unable to give an exact formula for $f(n)$ and are far from being able to determine the structure of the extremal graphs, i.e. we do not know the structure of the graphs $G(n; f(n) - 1)$ which do not contain a rectangle.

Let $a_1 < \ldots < a_k \leq x$ and assume that the product of any $r \, a_i$'s are different (or that the product of any r or fewer a_i's are different). I am not able to give a very satisfactory estimation for $\max k - \Pi(x)$ if $r > 2$. Perhaps the answer depends essentially on the fact whether we only require that the product of r or fewer distinct a_i's are all different or whether we permit repetitions.

Here we only state one result: Let $a_1 < \ldots < a_k \leq x$ be such that all products $\prod_{i=1}^{k} a_i^{\varepsilon_i}$, $\varepsilon_i = 0$ or 1, are distinct. Then [6]

(6) $$\max k < \Pi(x) + c_6 \, x^{1/2}/\log x.$$

The proof of (6) is not graph theoretical and will not be discussed here. Perhaps (6) can be improved to

(7) $$\max k < \Pi(x) + \Pi(x^{1/2}) + o(\frac{x^{1/2}}{\log x}) = \Pi(x) + \frac{(2 + o(1))x^{1/2}}{\log x}.$$

The inequality (7), if true, is best possible. To see this, let the a_i's be the primes and their squares.

An old and difficult conjecture of Turán and myself can be stated as follows: Let $a_1 < \ldots$ be an infinite sequence of integers and denote by $f(n)$ the number of solutions of $n = a_i + a_j$. Then $f(n) > 0$ for $n > n_o$ implies $\lim_m \sup f(n) = \infty$. A more general conjecture which is perhaps more amenable to attack goes as follows: Let $a_k < c \, k^2$, then $\lim_n \sup f(n) = \infty$. I could only prove that $a_k < c \, k^2$ implies that the sums $a_i + a_j$ cannot all be different [14]. We come to very interesting problems if we restrict ourselves to finite sequences. Let $A(n, r)$ be the largest integer so that there is a sequence $a_1 < \ldots < a_k \leq n$, $k = A(n, r)$ for which all sums of r or fewer a_i's are distinct. It is known that [7]

(8) $$(1 + o(1))n^{1/2} < A(n, 2) < n^{1/2} + n^{1/4} + 1.$$

I conjecture that $A(n, 2) = n^{1/2} + 0(1)$. Bose and Chowla proved [1]

$$A(n, r) \geq (1 + o(1))n^{1/r}$$

and they conjectured $A(n, r) = (1 + o(1))n^{1/r}$.

Let $a_1 < \ldots < a_k \leq n$ be a sequence of integers so that all the sums $\sum_{i=1}^{k} \varepsilon_i a_i$, $\varepsilon_i = 0$ or 1, are distinct. An old conjecture of mine states that

$$\max k = \frac{\log n}{\log 2} + 0(1).$$

Moser and I [8] proved

$$\max k \leq \frac{\log x}{\log 2} + \frac{\log\log x}{2 \log 2} + 0(1).$$

Conway and Guy proved (unpublished) that if $n = 2^r$ is sufficiently large then $\max 2^r \geq r + 2$.

These problems perhaps have nothing to do with graph theory, but often their multiplicative analogue can be settled by graph theoretic methods. In fact I proved the following theorem [9]. Let $a_1 < \ldots$ be an infinite sequence of integers. Denote by $g(n)$ the number of solutions of $n = a_i a_j$. Then if for $n > n_0$, $g(n) > 0$ we have $\lim_n \sup g(n) = \infty$, and in fact $g(n) > (\log n)^{c_7}$ for infinitely many n. This latest result cannot be improved very much since it fails to hold if c_7 is replaced by a sufficiently large constant c_8.

Denote by $u_p(n)$ the smallest integer so that if $a_1 < \ldots < a_k \leq n$, $k = u_p(n)$, is any sequence of integers then for some m, $g(m) \geq p$. We have for $2^{r-1} < p \leq 2^r$ [9],

$$(9) \qquad u_p(n) = (1 + o(1)) \, n(\log\log n)^{r-1}/(r-1)! \, \log n$$

$$= (1 + o(1)) \, \Pi_r(n) \, ,$$

where $\Pi_r(n)$ denotes the number of integers not exceeding n having r distinct prime factors.

For $p > 2$ I cannot at present get a result which is as sharp as (4). I just want to state without proof a special result in this direction, namely

$$(10) \quad \frac{n\log\log n}{\log n} + c_9 \, n/(\log n)^2 < u_3(n) < \frac{n\log\log n}{\log n} + c_{10} \, n/(\log n)^2.$$

It is not clear whether (10) can be sharpened.

The basic lemma needed for the proof of all these theorems is the following result on r-graphs: To every k and r there is an $\varepsilon_{k,r}$ so that every $G_r(n; c_{11} \, n^{r-\varepsilon_{k,r}})$ contains a $K_r(k, \ldots, k)$. For $r = 2$, $k = 2$, (5) shows that $\varepsilon_{k,r} = 1/2$. A result of Kövári, V.T. Sós and Turán [13] shows that $\varepsilon_{k,r} \geq 1/k$. In fact probably $\varepsilon_{k,2} = 1/k$ is the best value for $\varepsilon_{k,2}$. For $k = 3$ this is a result of W. Brown [2], but the cases $k > 3$ are still open. For $r > 3$ the best values of $\varepsilon_{k,r}$ are not known.

These extremal problems for r-graphs are usually much simpler for $r = 2$ (i.e. for the ordinary graphs). To illustrate this difficulty denote by $f(n, r, s)$ the smallest integer for which every $G_r(n; f(n, r, s))$ contains a $K_r(s)$. Turán determined $f(n, 2, s)$ for every n and s (e.g. $f(n, 2, 3) = [\frac{n^2}{4}] + 1$) and he posed the problem for $r > 2$ but as far as I know there are only inequalities and conjectures for $r > 2$. Turán conjectured that $f(2n, 3, 5) = n^2(n - 1) + 1$. It is easy to show that

$$\lim_{n \to \infty} f(n, r, s)/n^r = \delta_{r,s}$$

always exists and Turán proved $\delta_{2,s} = 1/2 - 1/2s$, but the value of

$\delta_{r,s}$ is unknown for every $s > r > 2$.

I would like to state one further conjecture for r-graphs: Every $G_3(3n; n^3 + 1)$ contains either a $G_3(4;3)$ or a $G_3(5;7)$.

Now I state a problem in number theory which can be reduced to a combinatorial problem:

Denote by $f(r, n)$ the smallest integer so that if $a_1 < \ldots < a_k \leq n$, $k = f(r, n)$ then there are r a_j's which pairwise have the same greatest common divisor. Using a combinatorial result of Rado and myself [11], I proved [12] that for every fixed r

(11) $$e^{c_r \log n/\log\log n} < f(r, n) < n^{3/4+\varepsilon}.$$

It seems that the lower bound in (10) gives the correct order of magnitude. This would follow (11) from the following conjecture of Rado and myself: There is a constant α_r so that if $A_1, \ldots A_s$, $s > \alpha_r^k$, are sets all having k elements, then there are always r of them, $A_{i_1}, \ldots, A_{i_r}$ which pairwise have the same intersection.

Finally, I would like to mention a few problems in combinatorial number theory: Let $a_1 < \ldots$ be an infinite sequence of integers, and assume that if

(12) $$\prod_{r=1}^{q_1} a_{i_r} = \prod_{r=1}^{q_2} a_{j_r}, \quad \text{then} \quad q_1 = q_2.$$

Is it true that for ε, there exists such a sequence of density $> 1 - \varepsilon$? Trivially, the a_i's can have density $1/4$. To see this, let the a_i's be the integers $\equiv 2 \pmod 4$. Selfridge showed that to every ε there is a sequence of density $> 1/e - \varepsilon$ satisfying (12). To see this let A be large and $A < p_1 < \ldots < p_k$ the sequence of consecutive primes satisfying

$$\sum_{i=1}^{k} 1/p_i < 1 < \sum_{i=1}^{k+1} 1/p_i.$$

The a_i's are the integers divisible by precisely one of the p_i's, $1 \leq i \leq k$. It is easy to see that for sufficiently large A, the a_i's have the required properties.

We come to non-trivial questions if we restrict ourselves to finite sequences. Let $a_1 < \ldots < a_k \leq n$ be a sequence of integers satisfying (12). How large can max k be? Is it true that max k = $n + o(n)$? I have no good upper or lower bounds for k. Trivially, max k $> n(\log 2 - o(1))$. To see this, consider the integers not exceeding n having a prime factor $> \sqrt{n}$. I can slightly improve the constant $\log 2$ but cannot prove max k = $n + o(n)$.

Let $a_1 < \ldots < a_k \leq n$; $b_1 < \ldots < b_q \leq n$ be two sequences of integers and assume that the products $a_i b_j$ are all distinct. Is it true that $kq < c \, n^2/\log n$?

Finally many of these problems can be modified as follows: Let $a_1 < \ldots < a_k$ be a sequence of real numbers. Assume that any two of the numbers $\Pi a_i^{\alpha_i}$ differ by at least one. Is it true that $\max k = \Pi(n)$?

REFERENCES

1. R.C. Bose and S. Chowla, Theorems in the additive theory of numbers, Comm. Math. Helv. 37 (1962-63), 141-147.

2. W.G. Brown, On graphs that do not contain a Thomsen graph, Canad. Math. Bull. 9 (1966), 281-285.

3. P. Erdös, On sequences of integers no one of which divides the product of two others, Izr. Inst. Math. and Mech. Univ. Tomsk 2 (1938), 74-82.

4. P. Erdös, On some applications of graph theory to number theoretic problems, Publ. Ramanujan Inst. (to appear).

5. P. Erdös, A. Rényi, and V.T. Sós, On a problem of graph theory, Studia Sci. Math. Hung. 1 (1966), 215-235.

6. P. Erdös, Extremal problems in number theory II, Mat. Lapok. 17 (1966), 135-155.

7. P. Erdös and P. Turán, On a problem of Lidon in additive number theory and on related problems, J. London Math. Soc. 16 (1941), 212-216.

8. P. Erdös, Problems and results in additive number theory, Coll. Théorie des Nombres, Brussels (1955), pp. 127-137.

9. P. Erdös, On the multiplicative representation of integers, Israel J. Math. 2 (1964), 251-261

10. P. Erdös, On extremal problems of graphs and generalized graphs, Israel J. Math. 2 (1964), 183-190.

11. P. Erdös and R. Rado, Intersection theorems for systems of sets, J. London Math. Soc. 35 (1960), 85-90.

12. P. Erdös, On a problem in elementary number theory and a combinatorial problem, Math. of Computation 18 (1964), 644-646.

13. T. Kövári, V.T. Sós, and P. Turán, On a problem of K. Zarankiewicz, Colloq. Math. 3 (1955), 50-57.

14. A. Stöhr, Gelöste und ungelöste Fragen über Basen der natürlichen Zahlenreihe I. 194 (1955), 40-65; II 194 (1955), 111-140.

ON THE NUMBER OF CYCLES
IN PERMUTATION GRAPHS

Joseph B. Frechen, St. John's University

In 1967 Chartrand and Harary introduced the concept of permutation graphs [1]. In that paper the authors characterize planar permutation graphs of 2-connected graphs. In [2] Hedetniemi generalized that concept to function graphs. Thus, the class of permutation graphs may now be studied from two points of view: the local point of view in which the internal structure of these graphs is examined; and the global viewpoint in which the common properties of this class and related classes of graphs are studied and compared.

It is the purpose of this paper to describe some aspects of the cycle structure in permutation graphs; the point of view will be the local one.

Let G be a graph with p points which are labeled $1, 2, \ldots, p$ and let α be a permutation on the set $\{1, 2, \ldots, p\}$. Then the $\underline{\alpha}$-permutation graph $P_\alpha(G)$ is defined to be the graph which consists of two disjoint, identically labeled copies of G, say G and G', together with p additional permutation lines which join G and G'. A permutation line $(j, \alpha(j))$ joins the point j in G with the point $\alpha(j)$ in G', $1 \leq j \leq p$. A graph H with $2p$ points is a permutation graph if there exist a graph G and a permutation α, both as described above, such that H is isomorphic to $P_\alpha(G)$. In a permutation graph $P_\alpha(G)$ it is often convenient to place primes on the labels of G' to avoid possible confusion of the points of G' with the points of G.

The first theorem relates the number of triangles and the number of 4-cycles of a graph G to the numbers of such cycles in the permutation graphs of G. In the statement and proof, $c(H)$ denotes the number of 4-cycles of a graph H.

Theorem 1. Let G be a graph with p points, q lines, t triangles and n 4-cycles. Then, for every permutation α,

    (a) $P_\alpha(G)$ has $2t$ triangles, and

    (b) $2n \leq c(P_\alpha(G)) \leq 2n + q$.

Moreover, $c(P_\alpha(G)) = 2n$ if and only if G is embeddable in its complement $\overline{G}$ where α is an appropriate isomorphism and $c(P_\alpha G)) = 2n + q$ if and only if α is an automorphism of G.

Proof. (a) Since each of the subgraphs G and G' of $P_\alpha(G)$ has t triangles, there are at least $2t$ triangles in every permutation graph of G. To see that there are no other triangles in $P_\alpha(G)$ for any permutation α, we examine the three vertices, say u, v and w, of a triangle T appearing in the permutation graph. If T is in G or in G', then it is already included in the count of 2t triangles. Let us suppose, then, that all of T is neither in G nor in G', so that one vertex of the triangle is in one of the subgraphs and the other two vertices are in the other subgraph. We may assume that u is in G and that v and w are in G'. But then the two sides of T which are adjacent at u are permutation lines of $P_\alpha(G)$. Since two permutation lines are never adjacent in a permutation graph, the supposed position of T in $P_\alpha(G)$ cannot occur.

(b) We may assert that $c(P_\alpha(G))$ is at least 2n for any permutation α because $c(G) = c(G') = n$, by hypothesis. None of the 2n 4-cycles in the subgraphs G and G' joins a point in G with a point in G'. Additional 4-cycles of $P_\alpha(G)$, when there are any, contain at least one permutation line. Let C be such a 4-cycle; then two of its four vertices must be in G and two in G'. For to have just one vertex of C, say u, in G and the other three vertices in G' implies that two permutation lines are adjacent at u. A new 4-cycle C is added to the base number of 2n each time two adjacent points u and v in G are joined by permutation lines to the points $\alpha(u)$ and $\alpha(v)$ in G' and the latter two points are adjacent also.

If G is such that $c(P_\alpha(G)) = 2n$ for some permutation α, then for each line (u,v) of G, $(\alpha(u),\alpha(v))$ is not a line. Thus if one considers α as a mapping from V(G) to $V(\overline{G})$, then each line of G is mapped into a line of $\overline{G}$ so that $\overline{G}$ contains G as a subgraph. The converse follows also since this argument is reversible.

The maximum number of 4-cycles exists in $P_\alpha(G)$ if the permutation α is such that whenever (u,v) is a line in G then $(\alpha(u), \alpha(v))$ is a line in G'. In this case, exactly q new 4-cycles are added. This implies that α is an automorphism of G. Conversely, if α is an automorphism of G, then $c(P_\alpha(G)) = 2n + q$.

One class of graphs for which the lower bound is attained is the family of self-complementary graphs. This is illustrated in Figure 1 where the graph G has its automorphism group $\Gamma(G)$ of order two, the identity being denoted by ϵ.

G: 1   2   3   4   n = c(G) = 0

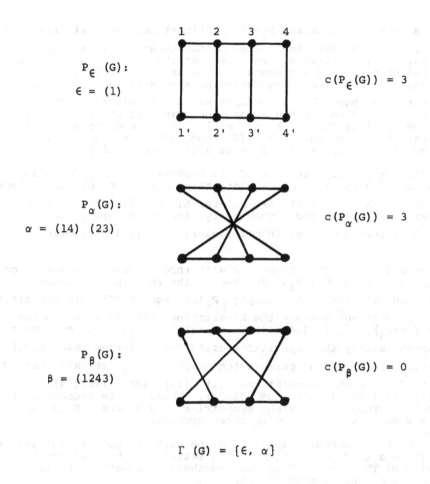

$$\Gamma (G) = \{\epsilon, \ \alpha\}$$

Figure 1. An example which illustrates the lower
bound of Theorem 1.

The next theorem characterizes bipartite permutation graphs. By
definition, a graph G is bipartite if its point set V(G) can be
partitioned into two disjoint point sets X and Y such that every
line of G has one end point in X and the other end point in Y.
For such a graph it will be convenient to refer to the pair {X,Y}
as a bipartition of V(G).

Theorem 2. Let G be a nonempty graph with p points. An α-per-
mutation graph of G is bipartite if and only if G is bipartite
and the permutation α is such that {α(X), α(Y)} is a bipartition
of V(G') whenever {X,Y} is a bipartition of V(G).

Proof. Necessity. Suppose $P_\alpha(G)$ is bipartite. Then all its cycles are even; in particular, the cycles of the subgraph G are even. Therefore, G is bipartite. To show that the permutation α has the stated property, we introduce a black and red 2-coloring of $P_\alpha(G)$ and let this bicoloring determine a bipartition $\{X,Y\}$ of $V(G)$. Let us suppose that the points of X are black and those of Y are red. Since each permutation line joins two points of different colors, we see that the points of $\alpha(X)$ are red and those of $\alpha(Y)$ are black. Also, the union of $\alpha(X)$ and $\alpha(Y)$ is $V(G')$. Therefore, the pair $\{\alpha(X), \alpha(Y)\}$ is a bipartition of $V(G')$.

Sufficiency. Suppose that G is bipartite and a permutation α has the stated property. We 2-color the points of $P_\alpha(G)$ as follows: the points of X and $\alpha(Y)$ are colored black while the points of Y and $\alpha(X)$ are colored red. Then every line in $P_\alpha(G)$ joins two points of different colors; in particular, this is a 2-coloring of $P_\alpha(G)$.

Figure 2 illustrates Theorem 2 with three permutation graphs of the bipartite graph $G = P_4$. For two of the three permutations, $\epsilon = (1)$ and $\alpha = (24)$, the graphs $P_\epsilon(G)$ and $P_\alpha(G)$ are bipartite, because β does not satisfy the bipartition criterion. We observe that $\alpha \notin \Gamma(P_4) = \{(1), (14)(23)\}$, so that the class H of permutations which satisfy the bipartition criterion is larger than $\Gamma(P_4)$. Examination of all of the permutation graphs of P_4 reveals that H consists of the eight permutations (1), (13), (24), $(12)(34)$, $(13)(24)$, $(14)(23)$, (1234) and (1432). Thus, H is isomorphic to the dihedral group of the eight symmetries of a square. This example and others suggest the following three problems.

1. Let G be a nonempty bipartite graph with p points, and let H be the class of permutations which satisfy the bipartition criterion of Theorem 2. Determine whether, in general, H is a subgroup of the symmetric group S_p.

2. Find the bipartite graphs G for which $H = \Gamma(G)$.

3. A generalization of problems 1 and 2.
   Consider the class $H(P,G)$ of all permutations α which preserve a property P of a graph G under transformation from G to $P_\alpha(G)$ (i.e., $P_\alpha(G)$ also possesses property P). Determine the properties P and graphs G for which
   (a) $H(P,G)$ is a subgroup of S_n, $n = |V(G)|$;
   (b) $H(P,G) = \Gamma(G)$.

$P_4:$

$P_\epsilon (P_4):$

$\epsilon = (1)$

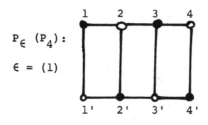

$P_\alpha (P_4):$

$\alpha = (24)$

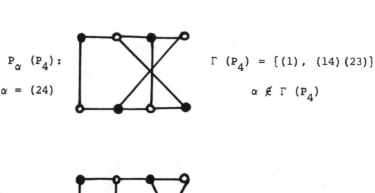

$\Gamma (P_4) = \{(1), (14)(23)\}$

$\alpha \notin \Gamma (P_4)$

$P_\beta (P_4):$

$\beta = (34)$

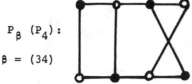

Figure 2. An example which illustrates the
bipartition criterion of Theorem 2.

REFERENCES

1. G. Chartrand and F. Harary, Planar permutation graphs, _Ann. Inst. Henri Poincaré_, Vol. IIIB (1967), 433–438.

2. S. Hedetniemi, On classes of graphs defined by special cutsets of lines, (this volume).

A NOTE ON A CATEGORY OF GRAPHS

Dennis P. Geller, University of Michigan
and Stephen T. Hedetniemi, University of Iowa

A theorem that is folklore not only in graph theory but in many algebraic systems is the following: An object G (graph, group, etc.) is isomorphic to a subobject of a direct product $G_1 \times G_2$ if and only if there exist homomorphisms $\phi_i: G \to G_i$ such that $\phi_1^{-1} \cap \phi_2^{-1} = 1$, the identity partition on G. In particular, this is a theorem in graph theory when direct product is taken to be the conjunction operation [1], also called Kronecker product [3], and homomorphisms are maps which preserve adjacency (see [2, p. 83]). It is our object in this note to point out that such a theorem fails to hold for a familiar class of mappings.

A <u>contraction</u> of a graph G to a graph G' is a map $\theta: V(G) \to V(G')$ such that for every point $v' \in V(G')$ the inverse image $\theta^{-1}(v')$ induces a connected subgraph of G. A binary operation $*$ defined on graphs will be called <u>contractive</u> if the graph G is a subgraph of the product of graphs $G_1 * G_2 * \ldots * G_n$ if and only if there are n contractions $\theta_i: G \to G_i$ such that $\cap \theta_i^{-1}$ is the identity partition of G.

<u>Theorem</u>: There exists no contractive binary operation on graphs.

<u>Proof</u>: The proof relies on the simple observation that a contraction of a graph will not, in general, be a contraction of its subgraphs. To utilize this fact we define a <u>coding</u> of a connected graph G to be a set of n contractions $\theta_i: G \to K_2$ such that for any $u, v \in V(G)$ the words $\theta_1(u)\,\theta_2(u)\,\ldots\,\theta_n(u)$ and $\theta_1(v)\,\theta_2(v)\,\ldots\,\theta_n(v)$ are distinct. If G is connected, let $\eta(G)$ be the smallest number of contractions in any coding of G. It is not difficult to show that any connected graph G has a coding, and that $\eta(G) = \Sigma\eta(B)$ where the sum is taken over all blocks B of G. In particular, a tree T with p points has $n(T) = p - 1$, while for the nontrivial complete graphs, $\eta(K_p) = \{\log_2 p\}$.

Now, suppose that $*$ is a contractive binary operation. It is clear that for any connected graph G, $G \leq K_2 * K_2 * \ldots * K_2$ where the product has $\eta(G)$ factors. But then if H is a connected subgraph of G we must have $\eta(H) \leq \eta(G)$. However for the path of length 7, $\eta(P_7) = 7$ even though $P_7 \subseteq K_8$ and $\eta(K_8) = 3$. This is a contradiction, and hence no contractive binary operation is possible on graphs.

REFERENCES

1. F. Harary and G. Wilcox, Boolean operations on graphs, _Math. Scand._, 20 (1967), 41-51.

2. O. Ore, _Theory of Graphs_, Amer. Math. Soc. Colloq. Publ., Vol. 38, Providence, 1962.

3. P.M. Weichsal, The Kronecker product of graphs. _Proc. Amer. Math. Soc._, 13 (1932), 47-52.

RECONSTRUCTING GRAPHS

D.L. Greenwell and R.L. Hemminger,[1]
Vanderbilt University

For any vertex v in $V(G)$ let G_v be the subgraph of G with $V(G_v) = V(G) - \{v\}$ and $E(G_v) = E(G) - \{e: e$ is incident to v in $G\}$. The following conjecture was proposed by P.J. Kelly [9] and S.M. Ulam [12].

<u>The Vertex Problem</u>. If G and H are graphs, $|V(G)| > 2$, and $\sigma: V(G) \rightarrow V(H)$ is a one-to-one onto function such that $G_v = H_{\sigma(v)}$ for all v in $V(G)$, then $G = H$.

An equivalent formulation of this problem is as follows: G can be uniquely determined, up to isomorphism, by the collection $\{G_v | v \in V(G)\}$. If this can be done we say that G can be reconstructed from that collection of subgraphs.

If $V(G) = \{v,w\}$ and $G_v = G_w = K_1$, then G cannot be reconstructed from G_v and G_w since K_2 and $K_1 + K_1$ yield these subgraphs. However, there are no other known couter-examples.

In [9], P.J. Kelly proved the vertex problem true for trees. There have also been a number of improvements on this result which we will discuss.

For any element e of $E(G)$ let G^e be the subgraph of G with $V(G^e) = V(G)$ and $E(G^e) = E(G) - \{e\}$. A problem similar to the vertex problem is the following:

<u>The Edge Problem</u>. If G and H are graphs, $|E(G)| > 3$, and $\sigma: E(G) \rightarrow E(H)$ is a one-to-one onto function such that $G^e = H^{\sigma(e)}$ for all e in $E(G)$, then $G = H$. Or equivalently: G can be reconstructed from the collection $\{G^e | e \in E(G)\}$.

We will see that the edge problem is a special case of the vertex problem.

If $E(G) = \{e,f,g\}$ and $G^e = G^f = G^g = K_1 \cup K(1,2)$ then G cannot be reconstructed from the collection $\{G^e | e \in E(G)\}$ since $K_1 + K_3$ and a 3-star yield the same subgraphs. Also if $E(G) = \{e,f\}$ and $G^e = G^f = K_1 + K_1 + K_2$, then G cannot be reconstructed since $K_2 + K_2$ and $K_1 \cup K(1,2)$ yield the same subgraphs. However,

---

[1]Research partially supported by NSF Grant #GY-4519.

for any graph G such that $|E(G)| > 3$ there are no known counter-examples.

The purpose of this paper is twofold. First we wish to summar-ize most of the known results and give some indication of the methods of their proofs. Secondly, we will include some new results. In particular we will show that the vertex and edge problems are true for a large class of graphs having terminal edges. Also most of the results on the edge problem are being published for the first time.

1. <u>Properties of G determined by the maximal subgraphs</u>. In solv-ing the vertex problem for a class of graphs it is often necessary to know properties of G that are determined by the maximal subgraphs. Lemma 1.1 is one of the most useful results in this direction. Lemma 1.2 gives the corresponding result for the edge problem.

<u>Lemma 1.1 (Kelly [9])</u>. If the conditions of the vertex problem are satisfied, then every type of vertex proper subgraph which occurs in G or H occurs the same number of times in both, and v and $\sigma(v)$ are vertices in the same number of these subgraphs, for all v in $V(G)$.

<u>Proof</u>. Let T denote a certain type of graph on j vertices, where $2 \le j < |V(G)|$, which occurs as a subgraph α times in G and β times in H. Also let $V(G) = \{v_i: i = 1,\ldots,n\}$ and α_i be the number of T-type subgraphs which has v_i as a vertex. Then

(1)
$$\alpha = \sum_{i=1}^{n} \alpha_i/j \quad \text{and} \quad \beta = \sum_{i=1}^{n} \beta_i/j$$

where β_i is the number of T-type subgraphs having $\sigma(v_i)$ as a ver-tex. Since $G_{v_i} = H_{\sigma(v_i)}$, the number of T-type subgraphs which do not have v_i as a vertex is the same as the number which do not have $\sigma(v_i)$ as a vertex. Thus, $\alpha - \alpha_i = \beta - \beta_i$, $i = 1,\ldots,n$. Therefore $\sum_{i=1}^{n} (\alpha - \beta) = \sum_{i=1}^{n} (\alpha_i - \beta_i)$ and hence $n(\alpha - \beta) = j(\alpha - \beta)$ from (1). But since $n \ne j$ we have $\alpha = \beta$. Since $\alpha = \beta$ we also know $\alpha_i = \beta_i$.

<u>Lemma 1.2</u>. If the conditions of the edge problem are satisfied then every type of edge proper subgraph which occurs in G or H occurs the same number of times in both, and e and $\sigma(e)$ are edges in the same number of these subgraphs, for all e in $E(G)$.

To illustrate the formulation of these problems in reconstruc-tion terms we give that version (with proof) of Lemma 1.2.

<u>Lemma 1.2'</u>. Let the family of graphs $\{G^e|\ e \in E(G)\}$ be given. Then the number of edge-proper subgraphs of G isomorphic to a given graph is determined by $\{G^e|\ e \in E(G)\}$ as well as the number of these that contain a given edge of G.

<u>Proof</u>. Let T be a graph with j edges, $1 \le j < |E(G)|$. Let α be the number of subgraphs of G that are isomorphic to T and α_i the number of subgraphs of G that are isomorphic to T and that contain the element e_i of $E(G)$, where $E(G) = \{e_i: i = 1,...,m\}$. Then $\alpha - \alpha_i$ is known since it is the number of subgraphs of G^{e_i} that are isomorphic to T. Thus from $\{G^e| e \in E(G)\}$ and T we know m, j, and $\alpha - \alpha_i$ for each $i = 1, 2, ..., m$. Therefore we know M where $M = \sum_{i=1}^{m} (\alpha - \alpha_i) = m\alpha - \sum_{i=1}^{m} \alpha_i = j\alpha$ so $M = (m - j)\alpha$. But $m \ne j$ so $\alpha = M/m - j$ is known and hence $\alpha_i = \alpha - (\alpha - \alpha_i)$ is known for each $i = 1, 2, ..., m$.

The <u>graph union</u> of two graphs G_1 and G_2 is the graph $G_1 \cup G_2$ with $V(G_1 \cup G_2) = V(G_1) \cup V(G_2)$ and $E(G_1 \cup G_2) = E(G_1) \cup E(G_2)$.

<u>Lemma 1.3 (Kelly [9])</u>. If the conditions of the vertex problem are satisfied then $\deg(v_i) = \deg(\sigma(v_i))$ for $i = 1, 2, ..., n$ and $|E(G)| = |E(H)|$. Equivalently we have that the degree sequence of G and the number of edges of G are determined by $\{G_v| v \in V(G)\}$.

<u>Proof</u>. In Lemma 1.1 let $T = K_2$ and we have $\deg(v_i) = \alpha_i = \beta_i = \deg(\sigma(v_i))$, and $|E(G)| = \alpha = \beta = |E(H)|$.

<u>Lemma 1.4</u>. The collection $\{G^e| e \in E(G)\}$ determines the degree sequence of G and $|V(G)|$ if $|E(G)| > 3$.

<u>Proof</u>. For each edge $e = (v_1, v_2)$ of G we can determine the degree of v_1 and v_2 in the following way:

By Lemma 1.2 we can determine the number of k-stars containing e if $k < |E(G)|$. Let N be the number of 2-stars containing e and let $p = |E(G)|$. Note that:

(1) $$N = \deg(v_1) + \deg(v_2) - 2$$

If the number of (p-1)-stars containing e is larger than one, with $p > 3$, then G must be a p-star and hence its degree sequence is determined.

If the number of (p-1)-stars containing e is one, then $\deg(v_1) = p - 1$ and, by (1), $\deg(v_2) = N - p + 3$, or vice versa.

If the number of (p-1)-stars is zero, then let $k + 1$ be the first integer such that the number of (k+1)-stars containing e is zero. Then if the number of k-stars containing e is two we have $\deg(v_1) = \deg(v_2) = k$. If the number of k-stars containing e is one we have $\deg(v_1) = k$ and $\deg(v_2) = N - k + 2$ or vice versa.

Since in this process each vertex of degree $k > 0$ is counted k times, the number of vertices of degree k in G is the number of vertices of degree k determined above, divided by k. Thus we have determined the number of vertices of each positive degree. But

$|V(G)| = |V(G^e)|$ for any $e \in E(G)$ so we also know the number of vertices of degree zero.

Lemma 1.4 is a little more interesting when viewed in terms of two graphs, and seems to say a little more; however, the two versions are equivalent.

Lemma 1.4'. If G and H are graphs, $|E(G)| > 3$, and there is a one-to-one onto function $\sigma: E(G) \to E(H)$ such that $G^e = H^{\sigma(e)}$, then for any $e = (v_1, v_2)$ in $E(G)$ with $\sigma(e) = (w_1, w_2)$ we have either $\deg(v_1) = \deg(w_1)$ and $\deg(v_2) = \deg(w_2)$ or else $\deg(v_1) = \deg(w_2)$ and $\deg(v_2) = \deg(w_1)$.

Corollary 1.5. The vertex and edge problems are true for k-regular graphs.

The graph G is <u>complement-connected</u> if both G and its complement $C(G)$ are connected.

Lemma 1.6. The vertex problem is true for the complement of G if and only if it is true for G.

Proof. Suppose there is a one-to-one onto function $\sigma: V(G) \to V(H)$ such that $G_v = H_{\sigma(v)}$ for all v in $V(G)$. Then $\sigma: V(C(G)) \to V(C(H))$ such that $C(G) = C(G_v) = C(H_{\sigma(v)}) = C(H)_{\sigma(v)}$. Since the vertex problem is true for the complement of G we have $C(G) = C(H)$. Therefore $G = H$. The converse follows immediately since $C(C(G)) = G$.

As seen in Corollary 1.5 the approach to the vertex or edge problem is to solve it for a class of graphs, hoping to eventually include every graph in one of the solved classes. Thus it is quite desirable to know if the maximal subgraphs of G determine whether G is in a given class or not. The following results are of that nature.

Lemma 1.7. The collection $\{G_v | v \in V(G)\}$, $|V(G)| > 2$, determines the connectivity of G.

Proof. If G is connected then not every vertex is a cut-point. So for some v in $V(G)$, G_v is connected and $|E(G_v)| < |E(G)|$. The converse is obvious. Hence, G is connected if and only if G_v is connected and $|E(G_v)| < |E(G)|$ for some v in $V(G)$.

Lemma 1.8. The collection $\{G^e | e \in E(G)\}$ determines the connectivity of G if $|E(G)| > 3$.

Proof. If $|E(G)| > 3$ and G^e is a star plus one isolated vertex for each e in $E(G)$, then G is a star and hence is connected.

If G^e is a forest with exactly two trees for all e in $E(G)$ and for some e neither component is an isolated vertex, then G is a tree and not a star.

If G^e is connected for some e in $E(G)$, then G is connected and not a tree.

Since the converse of these statements also hold, we have the lemma.

Theorem 1.9 (Harary [7]). If G is disconnected then the collection $\{G_v \mid v \in V(G)\}$ determines the connected components of G.

Proof. To emphasize the usefulness of Kelly's Lemma we give a different proof than that in [7].

Let G, H, and σ be as in the statement of the vertex problem, let B_i, $i = 1, 2, \ldots, m$ be the connected components of G, and let C_j, $j = 1, 2, \ldots, n$ be the connected components of H. Since $m \geq 2$ we have $n \geq 2$ by Lemma 1.7.

We say that a graph M has order (v,e), denoted by $o(M)$ if $v = |V(M)|$ and $e = |E(M)|$. Moreover if $o(M) = (v,e)$ and $o(M') = (v',e')$ we say $o(M) < o(M')$ if $v < v'$ or else $v = v'$ and $e < e'$.

Without loss of generality we assume that $o(B_1) \geq o(B_2) \geq \ldots \geq o(B_m)$ and $o(B_1) \geq o(C_1) \geq o(C_2) \geq \ldots \geq o(C_n)$. Now $|V(B_1)| < |V(G)|$ since G is not connected so, by Kelly's Lemma, H has a subgraph $H_1 = B_1$. Thus H_1 is connected so H_1 is a subgraph of C_j for some j. But also $o(H_1) = o(B_1) \geq o(C_1) \geq o(C_j)$ so $H_1 = C_j$. We can relabel so that $B_1 = C_1$.

Suppose that $B_i = C_i$ for $1 \leq i \leq k$ with $1 < k < m$, and that $o(B_{k+1}) \geq o(C_{k+1})$. As before, by Kelly's Lemma, G and H have the same number of subgraphs isomorphic to B_{k+1}. But $\bigcup_{i=1}^{k} B_i$ and $\bigcup_{i=1}^{k} C_i$ also have the same number of subgraphs isomorphic to B_{k+1} so H has a subgraph $H_{k+1} = B_{k+1}$ with H_{k+1} a subgraph of C_j for some $j > k$. Thus $H_{k+1} = C_j$ since $o(H_{k+1}) = o(B_{k+1}) \geq o(C_{k+1}) \geq o(C_j)$. We can relabel so that $B_{k+1} = C_{k+1}$.

The theorem follows by induction.

Lemma 1.10. If G has at least two non-trivial connected components then the collection $\{G^e \mid e \in E(G)\}$ determines the connected components of G.

Proof. The method of determining the connected components on two or more vertices is similar to the technique used in the proof of Theorem 1.9 except that Lemma 1.2 is used instead of Lemma 1.1. The lemma then follows immediately.

Using the same technique we get two more results. (A _leaf_ is a maximal connected subgraph without bridges.)

Lemma 1.11 (Bondy [2]). If G is connected, has cutpoints, and
$|V(G)| \geq 3$, then the collection $\{G_v | \ v \in V(G)\}$ determines the
blocks of G.

Lemma 1.12. If G is connected, has cut edges, and $|E(G)| > 3$ then
the collection $\{G^e | \ e \in E(G)\}$ determines the leaves of G.

Often one doesn't need all of the G_v to determine properties of
G. For example, if we let $S = \{v : |E(G_v)| = |E(G)| - 1\}$, then we
have

Lemma 1.13. If G is connected then the collection $\{G_v | \ v \in S\}$ de-
termines if G is a tree or not.

Proof. G is a tree if and only if G_v is a tree for all $v \in S$.

A vertex c is called a <u>center</u> of a tree T if for all v in
$V(T)$, $\max\{d(c,w) : w \in V(T)\} \leq \max\{\overline{d}(v,w) : w \in V(T)\}$. If T has
only one center it is said to be <u>centered</u>. If T has two centers it
is said to be <u>bicentered</u>. No tree has more than two centers. The
<u>radius</u> of T is $r = \max\{d(c,w) : c$ is a center and $w \in V(T)\}$. The
<u>diameter</u> of T is 2r if T is centered and $2r + 1$ if T is bi-
centered.

Lemma 1.14. If G is a tree the collection $\{G_v | \ v \in S\}$ determines
whether G is cenetered or bicentered.

Proof. The lemma is easily checked if $|S| < 5$ and if $|S| \geq 5$ then
G is centered if and only if at most two of the G_v, $v \in S$ are bicentered trees. For let $P(v_1, v_2)$ be a diameter of G. Then
$P(v_1, v_2)$ is a diameter of G_v for all $v \in S - \{v_1, v_2\}$. Thus the
center of $P(v_1, v_2)$ is the center of G_v for all $v \in S - \{v_1, v_2\}$.
The converse follows similarly if G, and hence $P(v_1, v_2)$, is bicentered.

There are other properties of G that are determined by the maximal subgraphs but many are trivial and the others are not used here
so we omit them.

2. <u>The edge problem</u>. A perusal of Section 1 suggests that there
might be some connection between the edge problem and the vertex problem. The purpose of this section is to show that in fact the edge
problem is a special case of the vertex problem. This result has appeared in [8]; however, since only the idea of the proof appeared
there, we will now give the details. We denote the line-graph of a
graph G by $L(G)$.

Theorem 2.1 (Whitney [13 or see pp. 248 of 11]): If G and H are
connected graphs other than triangles, then $G = H$ if and only if
$L(G) = L(H)$.

Lemma 2.2. Let G be a graph. Then $L(G^e) = (L(G))_e$ for all e in
$E(G)$.

Proof. We have $V(L(G^e)) = E(G) - \{e\} = V((L(G)_e)$ and (e_1, e_2) is in $E(L(G^e))$ if and only if $e_1, e_2 \neq e$ and e_1 and e_2 are adjacent in G, that is if and only if (e_1, e_2) is in $E((L(G))_e)$. Therefore $E(L(G^e)) = E((L(G))_e)$. Thus $L(G^e) = (L(G))_e$.

Theorem 2.3 (Hemminger [8]). The edge problem is true for G if and only if the vertex problem is true for $L(G)$.

Proof. Suppose the vertex problem is true for line graphs. Let G and H be graphs, $|E(G)| > 3$, and let $\sigma: E(G) \to E(H)$ be a one-to-one onto function such that $G^e = H^{\sigma(e)}$ for all e in $E(G)$. By Lemma 1.10 if G is disconnected, then $G = H$. So suppose G is connected. By Lemma 2.2 we have $(L(G))_e = L(G^e) = L(H^{\sigma(e)}) = (L(H))_{\sigma(e)}$ for all e in $E(G)$. But then $\sigma: V(L(G)) \to V(L(H))$ is a one-to-one onto function such that $(L(G))_e = (L(H))_{\sigma(e)}$ for all e in $V(L(G))$ and $|V(L(G))| > 2$. So by our assumption $L(G) = L(H)$. Since G and H are connected $L(G)$ and $L(H)$ are connected. Hence by Whitney's Theorem, since $|E(G)| > 3$, we have $G = H$.

Conversely, suppose the edge problem is true for graphs. Let G and H be graphs with $|V(L(G))| > 2$, and let $\sigma: V(L(G)) \to V(L(H))$ be a one-to-one onto function such that $L(G)_e = L(H)_{\sigma(e)}$ for all e in $V(L(G))$. If $L(G)$ is disconnected then by Theorem 1.9, $L(G) = L(H)$. So suppose that $L(G)$ (and hence $L(H)$) is connected. Then G and H are graphs with one and only one non-trivial component. Denote these components by $\overline{G}$ and $\overline{H}$.

By Lemma 2.2 we have $L(G^e) = L(H)_{\sigma(e)} = L(H^{\sigma(e)})$. Hence $\sigma: E(\overline{G}) \to E(\overline{H})$ is a one-to-one onto function such that $L(\overline{G}^e) = L(\overline{H}^{\sigma(e)})$ for all e in $E(\overline{G})$.

Case 1. Suppose $n = |V(\overline{G})| = |V(\overline{H})| - p$ where $p > 0$. We know then that $L(\overline{G}^e)$ has at most n complete subgraphs in the canonical decomposition of a line graph into edge disjoint complete subgraphs, the largest number occurring when e is not a terminal edge of $\overline{G}$. Similarly we know $L(\overline{H}^{\sigma(e)})$ can have no fewer than n complete subgraphs in its canonical decomposition, the smallest number occurring when $\sigma(e)$ is a terminal edge of $\overline{H}$. But since $L(\overline{G}^e) = L(\overline{H}^{\sigma(e)})$ for all e in $E(\overline{G})$, they must have the same number of these complete subgraphs for each e, namely n. Hence $\overline{G}$ has no terminal edges and $\overline{H}$ is an n-star. The only time this can happen and have $L(\overline{G}^e) = L(\overline{H}^{\sigma(e)})$ is if $\overline{G} = K_3$ and $\overline{H}$ is a 3-star. But in this case $L(G) = L(H) = K_3$.

Case 2. Suppose $|V(\overline{G})| = |V(\overline{H})|$ and for some e in $E(\overline{G})$ we have $\overline{G}^e = K_1 + K_3$, and $\overline{H}^{\sigma(e)}$ is a 3-star. Then there is only one place to replace e in $\overline{G}^e$ and $\sigma(e)$ in $\overline{H}^{\sigma(e)}$ and both yield a graph isomorphic to K where $V(K) = \{a,b,c,d\}$ and $E(K) = \{(a,c), (b,c), (d,c), (b,d)\}$. So $L(G) = L(H)$.

Case 3. Suppose $|V(\overline{G})| = |V(\overline{H})|$ and for all e in $E(\overline{G})$ the sub-graphs above do not occur. Let e be any edge of $\overline{G}$. Since $L(\overline{G}^e)$ = $L(\overline{H}^{\sigma(e)})$, Whitney's Theorem tells us that the non-trivial components of $\overline{G}^e$ are isomorphic to the non-trivial components of $\overline{H}^{\sigma(e)}$ except for the triangle and 3-star components. But since $\overline{G}^e$ and $\overline{H}^{\sigma(e)}$ are not those subgraphs of case 2, $|V(\overline{G})| = |V(\overline{H})|$, and $\overline{G}^e$ and $\overline{H}^{\sigma(e)}$ each have at most two components, we must have the same number of non-trivial, triangle, and 3-star components in $\overline{G}^e$ as in $\overline{H}^{\sigma(e)}$. Hence $\overline{G}^e = \overline{H}^{\sigma(e)}$ for all e in $E(\overline{G})$. So by our assumption $\overline{G} = \overline{H}$, and thus $L(G) = L(H)$.

3. Reconstructible classes of graphs. We will now give most of the graphs for which the vertex or edge problem has been solved.

Theorem 3.1 (Harary [7]). The vertex problem is true for graphs that are not complement-connected, i.e. if G is not complement-connected then G can be reconstructed from $\{G_v | v \in V(G)\}$.

Proof. This is immediate from Lemma 1.6 and Theorem 1.9.

Theorem 3.2. The edge problem is true for disconnected graphs having at least two non-trivial connected components.

Proof. This is immediate from Lemma 1.10.

   Because of these two theorems we hereafter only consider connected graphs.

Theorem 3.3. The vertex problem is true for regular graphs.

Proof. This is Corollary 1.5.

Theorem 3.4. The edge problem is true for regular graphs.

Proof. This is Corollary 1.5.

Theorem 3.5 (Bondy [2]). The vertex problem is true for graphs with cutpoints but without terminal vertices (a terminal vertex is one of degree one).

Proof. Let B_i, i = 1, 2, ..., m be the "terminal" blocks of G and let C_j, j = 1, 2, ..., n be the "terminal" blocks of H where G, H, and σ are as in the statement of the vertex problem (a terminal block is one containing only one cut-point of G). Without loss of generality we assume that $(b_1, e_1) = o(B_1) \leq o(B_i)$ i = 2, 3, ... m and $o(B_1) \leq o(C_j)$, j = 1, 2, ..., n. Let u be the cut-point of G contained in B_1, let $G_1 = G_{V(B_1)} - u$, and let G_1^s be the graph obtained from G_1 by adding s isolated vertices and joining each to u by an edge. Then G_1^1 is a proper subgraph of G and hence, by Kelly's Lemma, H has a subgraph $H_1^1 = G_1^1$, say

$\psi: G_1^{\ 1} = H_1^{\ 1}$, that is ψ is an isomorphism from $G_1^{\ 1}$ to $H_1^{\ 1}$. Let $v = \psi(u)$. (Note: $\sigma(u)$ need not be equal to v). Let p be the terminal vertex of $H_1^{\ 1}$ and let H_1 be the subgraph of H obtained by deleting p from $H_1^{\ 1}$. Then it is clear that $\psi | G_1: G_1 = H_1$. Thus, by Lemma 1.11, H_1 has one block fewer than H. If we call that block C_1 then $C_1 = B_1$. Now H is obtained from $H_1^{\ 1}$ by adding $b_1 - 2$ vertices and some edges. Since no terminal block of H has order less than that of B_1 and since H has no terminal vertices it is easy to see that those edges can only be incident with v, p and the $b_1 - 2$ vertices added to $H_1^{\ 1}$. Thus v is a cutpoint of H and it follows that the subgraph of H on v, p, and the $b_1 - 2$ vertices added to $H_1^{\ 1}$ is isomorphic to C_1. It remains to show that there is an isomorphism of B_1 and C_1 that maps u to v. To do this we will show that $B_1^{\ 1} = C_1^{\ 1}$ where $B_1^{\ 1}$ is obtained from B_1 by adding a terminal edge at u and $C_1^{\ 1}$ from C_1 by adding one at v. Obviously $G_u = H_v$ so u and v have the same degree. Let $\deg(u) = r + s$ where the degree of u in G_1 is r and in B_1 is s. Hence $r, s > 0$. By the definition of H_1 it is clear that the degree of v in H_1 is r; hence the degree of v in C_1 is s. If $G_1^{\ s}$ has α subgraphs isomorphic to $B_1^{\ 1}$ then G has $\alpha + r$ subgraphs isomorphic to $B_1^{\ 1}$ since B_1 is a block in G. Thus H has $\alpha + r$ subgraphs isomorphic to $B_1^{\ 1}$ by Kelly's Lemma and $H_1^{\ s}$ has α subgraphs isomorphic to $B_1^{\ 1}$ since $H_1^{\ s} = G_1^{\ s}$. It follows, since C_1 is a block of H, that $C_1^{\ 1}$ has r subgraphs isomorphic to $B_1^{\ 1}$ and hence that $B_1^{\ 1} = C_1^{\ 1}$ since $B_1 = C_1$. That completes the proof of the theorem.

If P is a path of length n from a to b in G such that $\deg(a) = 1$, $\deg(b) \geq 3$, and $\deg(v) = 2$ for all vertices v in P other than a and b we call P a <u>twig</u> of G having length n.

<u>Theorem 3.6</u>. The edge problem is true for connected graphs with bridges but without twigs of length more than one.

<u>Proof</u>. If G is such a graph then $L(G)$ has cut-points but has no terminal vertices. Thus by Theorem 3.5 the vertex problem is true for $L(G)$. But then by Theorem 2.3 the edge problem is true for G.

We should point out that $\{G^e | e \in E(G)\}$ determines G is G is a graph of the type mentioned in the theorem.

The first result on the vertex problem was given by Kelly in 1957 where he showed it was true for trees. Since then most of the work has been on improving this result by reducing the number of subgraphs needed to reconstruct a tree (It is only with Bondy's result

(Theorem 3.5) and our results in the next section that any significant progress has been made on the vertex problem itself). We now summarize the known results for trees and indicate the methods by giving an outline and portions of the proof of Theorem 3.8.

Theorem 3.7 (Kelly [9]). The vertex problem is true for trees.

Theorem 3.8 (Harary and Palmer [4]). A tree T can be reconstructed from the collection $\{T_v| \ v \in S\}$ where $S = \{v \in V(T): \deg(v) = 1\}$.

Corollary 3.9. A tree T can be reconstructed from the collection $\{T^e| \ e \in R\}$ where $R = \{e \in E(T): e$ a terminal edge in $T\}$.

Proof. This follows immediately from Theorem 3.8 since the collection $\{T_v| \ v \in S\}$ can be obtained from the collection $\{T^e| \ e \in R\}$ by deleting the one isolated vertex from each T^e.

Corollary 3.10. The vertex problem is true for the line-graph of a tree.

Proof. This follows immediately from Theorem 2.3 and Corollary 3.9.

Theorem 3.11 (Bondy [1]). A tree T can be reconstructed from the collection $\{T_v| \ v \in P\}$ where $P = \{v \in V(T)|v$ is the endpoint of a diameter of $T\}$.

Manvel has given a proof of Theorem 3.8 that he was able to modify to prove the following which is the only results of this type.

Theorem 3.12 (Manvel [10]). With two exceptions, a tree T can be reconstructed from a collection $\{T_v| \ v \in M\}$ where $T_v \neq T_u$ if $u, v \in M$ with $u \neq v$ but where for each terminal vertex v in T there is a $u \in M$ such that $T_v = T_u$. G, a path of length 3, and H, a 3-star, gives one exception. The other exception is when G and H have exactly one vertex of degree three (none of larger degrees) and G has one twig of length three, two of length one; while H has one of length one and two of length two.

Before discussing a proof of Theorem 3.8 we need some definitions.

A branch B of a center c of a tree T is a rooted subtree of T maximal with respect to the properties:
  (a) B is rooted at c
  (b) B contains only one vertex adjacent to c.
A radial branch is a branch B of a centered tree T such that the max $\{d(r,v): r$ is the root of B and $v \neq r$ is any vertex in B$\}$ is the radius of T. A path branch is a branch B of a tree T which is a path and is rooted at one of its endpoints.

Outline of the proof. If T has less than five terminal vertices then T must be homeomorphic to either a 1-star, a 3-star, a 4-star, or the non-star graph obtained from two distinct 3-stars by identifying an edge from each. By examining the maximal subtrees in these cases we see that they determine T. Thus we can assume T has at least five terminal vertices and so by Lemma 1.14 the number of the maximal subtrees that are centered determined whether T is centered or bicentered. The proof is completed by considering the following cases:

  I-a: Each maximal subtree of T is centered.
  I-b: Each maximal subtree of T is bicentered.
  II-a: Exactly one maximal subtree is centered.
  II-b: Exactly one maximal subtree is bicentered.
  III-a: Exactly two maximal subtrees are centered.
  III-b: Exactly two maximal subtrees are bicentered.

There are a lot of similarities in the proofs for the different cases. We prove two of the cases to illustrate the techniques and because we will use one of them later.

<u>Proof of I-a:</u> By Lemma 1.14 T is centered. Let $k \geq 5$ be the number of maximal subtrees. Let M be the largest number such that there is a branch in some T_j, $1 \leq j \leq k$, with M edges. Let B_i, $1 \leq i \leq s$, denote the types of branches with M edges which appear in some T_j. Let α_{ij} be the number of branches of type B_i in T_j where $1 \leq i \leq s$ and $1 \leq j \leq k$.

If there is an i such that $\alpha_{ij} = p$ for all j then there are $p + 1$ of the B_i branches in T and no others. Otherwise for each i there is a q and an r such that $\alpha_{iq} = \alpha_{ir} + 1$. In which case there are α_{iq} branches of type B_i in G. The branches of T with $M - 2$ or fewer edges are those which appear in T_r. Now let $F = \{j: T_j$ has one less than the known number of branches of type $B_1\}$ and let $Q = \{(B_1)_v: v$ is the element of $V(B_1)$, $\deg(v) = 1$ and v is not the root of $B\}$. If $|Q| = 1$, then the $(M-1)$-edges branches are those of T_j, for any j in F, minus one of the type isomorphic (preserving root) to the element of Q. Otherwise they are the $(M-1)$-edged branches which appear in all the T_j where j is in F, and the number of times each occurs in T is the smallest number of times it appears in a T_j, where again j is in F. Thus we have determined the rooted branches of T. We reconstruct T by identifying the roots of these branches to a single points.

<u>Proof of III-b:</u> By Lemma 1.14 T is centered. Let T_1 and T_2 be bicentered, and let c be the center of T.

(1) Suppose $\deg(c) = 2$ and for each $i \geq 3$, T_i has a branch at c with only one endpoint. Since T has five endpoints, one of the two branches at c must be a path branch of length r (the radius of T). For each $i \geq 3$ let w_i be the point of T_i with $\deg(w_i) > 2$ and minimum distance, $d(c,w_i)$, from c. Let $d = \min \{d(c,w_i): i \geq 3\}$. Then d is the distance in T from c to the nearest point of degree greater than two. Let T_1 be the bicentered subtree with centers c_1 and c_2 and a vertex u such

that $\deg(u) > 2$ and $d(u,c_2) = d - 1$. Then the second branch of T is the branch of T_1 at c_1 which contains c_2.

(2) Suppose $\deg(c) = 2$ and for some $i \geq 3$, T_i has more than one endpoint on each of its two branches. We can determine these branches as in Part I-a.

(3) Suppose $\deg(c) > 2$ and for all $j \geq 3$, both radial branches of T_j are paths of length r (the radius of T). Then in T_1 let c_1 be the center whose degree is greater than two. The non-radial branches of T are the branches of T_1 at c_1 minus a path branch of length r and a path branch of length $r - 1$. The two radial branches of T are path branches of length r.

(4) Suppose $\deg(c) > 2$ and for some $j \geq 3$ one of the radial branches of T_j is not a path branch. Then the branches can again be determined as in Part I-a.

We close this section with a result (that follows immediately from Theorem 3.8) which we need in the next section.

Corollary 3.13 (Harary and Palmer [4]). Let T be a rooted tree with root v. Let $v_1, v_2, \ldots, v_n$ be the vertices of T, other than v, that have degree one. Then T can be reconstructed from the collection $\{T_i \mid i = 1,2,\ldots,n\}$ with the root v specified in each $T_i = T_{v_i}$.

4. Reconstructing graphs that have cutpoints. In light of Theorems 3.1, 3.5, and 3.7, the graphs for which the vertex problem has not been solved can be broken into two major classes: (1) connected graphs with circuits and terminal vertices, and (2) connected graphs with circuits and without cut-points, i.e. blocks. At the time of writing this paper, the authors have no idea of how to approach the reconstruction of blocks (nor do they know of anyone with such an idea); however, in this section we wish to present a solution for a subclass of the class of graphs in (1) by using a technique that will apply to all graphs in class (1). Our success in using the technique was in being able to handle automorphisms of the "central" block if it was quasi-dihedral while our failure was in not being able to handle a general automorphism of a general "central" block. But we still hold high hopes for being able to solve the vertex problem for the class of graphs in (1) via this approach. In fact our solution, as in the refinements of Kelly's result on trees, only uses the G_v with $\deg(v) = 1$. Thus our solution is to a restricted vertex problem which we refer to as:

The Terminal-Vertex Problem. If G and H are connected graphs with terminal vertices $S(G)$ and $S(H)$ respectively $|S(G)| \geq 1$, and $\sigma: S(G) \to S(H)$ is a one-to-one onto function such that $G_v = H_{\sigma(v)}$ for all $v \in S(G)$, then $G = H$.

An equivalent formulation is as follows: G can be uniquely determined, up to isomorphism, by the collection $\{G_v|\ v \in S(G)\}$.

Since the G_v with $v \in S(G)$ can be distinguished from the G_v with $v \in V(G) - S(G)$ a solution to the terminal-vertex problem is also a solution to the vertex problem. Also as pointed out with trees a terminal-edge problem would be equivalent to the terminal-vertex problem.

Before getting to our main result we need some definitions and preliminary results.

Definitions. (a) Let G be a connected graph with circuits. A proper subgraph T of G is called a tree growth of G if it is maximal with respect to the conditions: (1) T is a tree, and (2) there is a subgraph G' of G and a vertex v of G such that $G = T \cup G'$ and $V(T) \cap V(G') = \{v\}$. Note that for a given tree growth T, the subgraph G' and the vertex v are unique. We call v the root of T and we call the intersection of all the G', as T varies over all tree growths of G, the pruned graph of G.

(b) The cut-point block graph of a graph G is the graph with $V(G) = \{x: x$ is a block of G or x is a cut-point of $G\}$ and $E(G) = \{(x,y): x$ is a block of G and y is a cut-point of G with $y \in V(x)\}$. Note that the cut-point block graph of a connected graph is a centered tree.

(c) The block or cut-point of G corresponding to the center of the cut-point block graph of the pruned graph of G is called the pruned center of G.

(d) Let G be a graph with a block D as its pruned center and let v be an element of D. The branch of G at v is the largest connected subgraph of G rooted at v that contains no other vertices of D.

(e) Let G be a graph with a cut-point c as its pruned center. The branches of G are the following:
(1) The tree growth of G rooted at c (if any)
(2) The subgraphs B of G rooted at c of the form $B' \cup T$ where (a) B' is the graph union in G of the vertices of a connected component of the graph that results from deleting the pruned center of G from the cut-point block graph of the pruned graph of G and (b) T is the graph union of all the tree growths rooted at some vertex of B'.

(f) Let $\Gamma(G)$ be the group of automorphisms of G. If $v \in V(G)$ then the transitivity class of v, denoted by $\bar{v}$, is the set $\{w \in V(G):$ there is a $\tau \in \Gamma(G)$ with $\tau(v) = w\}$. We say that G is transitive if $V(G)$ is a transitivity class. Note that $\bar{v} = \bar{w}$ if and only if $\bar{v} \cap \bar{w} \neq \phi$. We say that $\bar{v}$ is independently transitive with respect to $\bar{w}$ if for each $v_1, v_2 \in \bar{v}$ and each $w_1 \in \bar{w}$ there is a $\tau \in \Gamma(G)$ such that $\tau(w_1) = w_1$ and $\tau(v_1) = v_2$. We say $\bar{v}$ and $\bar{w}$ are independently transitive if they are independently transitive with respect to each other.

Conventions. In the remainder of this section let G be a connected graph with circuits and with $|S(G)| = m \geq 1$. Moreover let

$S(G) = \{1, 2, \ldots, m\}$. When we refer to a G_v with $v \in V(G)$ we will assume $v \in S(G)$ unless otherwise stated. Let g be the total number of edges that are in tree growths of G. When we say a branch is known we mean its root is also known. The following is obvious.

Theorem 4.1. If $m = g = 1$ then G can be reconstructed from $\{G_v | v \in S(G)\}$ if and only if G_1 is transitive.

Theorem 4.2. If $m = g = 2$ (say $v = 1$ and $w = 2$ in $S(G)$) then G can be reconstructed from $\{G_v | v \in S(G)\}$ if and only if either $G_v = G_w$ and $|\overline{w}'| = 1$ in $(G_v)_w$ (w' is the vertex in G_v that is adjacent to vertex w) or else $G_v \neq G_w$ and $\overline{v}'$ and $\overline{w}'$ are independently transitive in $(G_v)_w$ (Note that $(G_v)_w = (G_w)_v$).

Proof. Suppose $G_v = G_w$. If $|\overline{w}'| = 1$ in $(G_v)_w$, then $v' = w'$ and G is determined. If $|\overline{w}'| > 1$ let $w'' \in \overline{w}'$ with $w'' \neq w'$. Let H be obtained from G_v by adding v and the edge (v,w'') and let G be obtained from G_v by adding v and the edge (v,w'). Then $G \neq H$ but they satisfy the hypothesis of the terminal-vertex problem.

Suppose $G_v \neq G_w$. Then $\overline{v}'$ in $(G_w)_v \neq \overline{w}'$ in $(G_v)_w$. If $\overline{v}'$ and $\overline{w}'$ are independently transitive in $(G_v)_w$ then G is determined by G_v and G_w and is isomorphic to the graph H obtained from G_v by adding w and an edge (w'',w) for any $w'' \in \overline{w}'$. If $\overline{v}'$ and $\overline{w}'$ are not independently transitive in $(G_v)_w$ then we can assume there is a v'' and $v''' \in \overline{v}'$ with $v'' \neq v'''$ and a $w'' \in \overline{w}'$ such that if $\tau \in \Gamma((G_v)_w)$ with $\tau(v'') = v'''$ then $\tau(w'') \neq w''$. Let H be obtained from $(G_v)_w$ by adding v and w and the edges (v,v'') (w,w'') and let G be obtained from $(G_v)_w$ by adding v and w and the edges (v,v''') and (w,w''). Then G and H are counterexamples to the terminal-vertex problem.

Technically, in the above, v' was in $(G_w)_v$ and w' was in $(G_v)_w$ and we could not use them to compare $\overline{v}'$ and $\overline{w}'$ for while $(G_v)_w = (G_w)_v$ we could not tell where v' was in $(G_v)_w$. However, if $\tau: (G_w)_v = (G_v)_w$, then we can identify $\overline{v}'$ in $(G_v)_w$ as $\tau(\overline{v}')$, and it was only $\overline{v}'$ (or $\overline{w}'$) that we used, not v' itself.

Theorem 4.3. If $m = g = 3$ (say $v_1 = 1$, $v_2 = 2$, and $w = 3$ in $S(G)$) and G_w has only one tree growth (where we let v' and w' be the vertices adjacent to vertices of degree one in $G_1 = G_2$) then G can be reconstructed from $\{G_v | v \in S(G)\}$ if and only if there is a $\tau \in \Gamma(G')$, $G' = ((G_1)_2)_3$, such that $\tau(w') = v'$ and $\tau(v') = w'$.

Proof. Let H be obtained from G_1 by adding v_1 and the edge

(v_1,v') and let G be obtained from G_1 by adding v_1 and the edge (v_1,w'). Then it is clear that G can be reconstructed from $\{G_v|$ $v \in S(G)\}$ if and only if $G = H$. The theorem follows immediately.

Theorem 4.4. If $m = 2$ and $g = 3$ (say $v = 1$ and $w = 2$ in $S(G)$ where G_w has only one tree growth and where we let v' and w' be the vertices in G_v adjacent to vertices of degree one) then G can be reconstructed from $\{G_v| v \in S(G)\}$ if and only if there is a $\tau \in \Gamma(G')$, $G' = ((G_1)_2)_3$, such that $\tau(v') = w'$ and $\tau(w') = v'$.

The proof of this theorem is similar to that of the last theorem.

The above four theorems describe the only known counterexamples to the terminal-vertex problem. Figure 1 illustrates a typical counter-example described by Theorem 4.3.

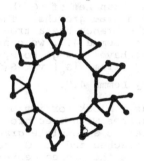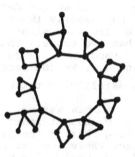

Figure 1

Lemma 4.5. If $m = 1$ and $g > 1$, or if $m > 1$ and $g > 2$ such that each G_v has only one tree growth, then G can be reconstructed from $\{G_v| v \in S(G)\}$.

Proof. Our assumptions guarantee us that G has only one tree growth T. By Corollary 3.13 we can reconstruct T from the T_v's where T_v is the tree growth of G_v. We reconstruct G from any G_v by replacing the T_v by T.

Lemma 4.6. If for some $v \in S(G)$, G_v has two branches with tree growths, then we can determine the branches of G.

The proof of Lemma 4.6 is similar to Part I-a of Theorem 3.8.

Lemma 4.7. If the branches of G are known and if for some branch B of G with more than one edge and some endpoint v in B, B_v is not a branch of G, then G can be reconstructed from $\{G_v| v \in S(G)\}$.

Proof. G is obtained from G_v by replacing the single branch of

G_v that is isomorphic to B_v by the branch B.

Lemma 4.8. If the branches of G are known and if the pruned center of G is a cut-point, then G can be reconstructed from $\{G_v |$ $v \in S(G)\}$.

Proof. We reconstruct G by identifying the roots of the branches of G to a single point.

Theorem 4.9. If G is not one of the known counterexamples, and if G has a cut-point as its pruned center or if G has at most two branches with tree growths then G can be reconstructed from $\{G_v |$ $v \in S(G)\}$.

Proof. Because of the hypothesis and the previous results we can assume that some G_v, and hence G, has at least two branches with tree growths. Thus we can determine the branches of G by Lemma 4.6. Suppose G has exactly two branches with tree growths. Then by Lemma 4.7, we can reconstruct G if some branch has a tree growth with more than two edges. Thus, again by Lemma 4.7, we can assume $g = 2$ or 3 and $m \geq 2$. But we always have $m \leq g$ so by Theorems 4.2, 4.3, and 4.4, we can reconstruct G from $\{G_v | v \in S(G)\}$. The remaining possiblity is taken care of by Lemma 4.8.

Thus we can now assume that G has a block for its pruned center and that G has at least three branches with tree growths. We note in passing that if the pruned center is a complete graph then it doesn't matter where the branches are attached and so G can be reconstructed. We would also point out that the pruned center of G is the same as the pruned center of each G_v for $v \in S(G)$. It is this fact that is the key to our approach in proving the following.

Theorem 4.10. If G has at least three branches with tree growths and if the pruned center of G is a circuit then G can be reconstructed from $\{G_v | v \in S(G)\}$.

Proof. As usual in the reconstruction problem there are a number of cases. Since they are all rather similar and all depend on the idea of a stacking (defined below) we will give the proof in only one case; namely, when there exists a branch B of G and $v, w \in V(B) \cap S(G)$, $v \neq w$, such that $B_v \neq B_w$.

We also note that by the hypothesis of the theorem we know the branches of G including their roots (by Lemma 4.6) and if B is a branch of G that has more than one edge and if there is a $v \in V(B)$ $\cap S(G)$ then we can assume, by Lemma 4.7, that there is a branch $B' = B_v$ (preserving root) in G.

Before continuing the proof we introduce, and prove a result about, stackings.

Definitions. A **pre-stacking** of $\{G_i | i = 1, 2, \ldots, m\}$ is a labeling of the vertices of the pruned centers C_i of G_i as $C_i = (a_{i0}, a_{i1}, \ldots, a_{ip})$ for $i = 1, 2, \ldots, m$ and $C_0 = (a_0, a_1, \ldots,$

a_p), where $G_0 = G$, so that $\sigma_i: C_i \to C_0$ given by $\sigma_i(a_{ik}) = a_k$ is an isomorphism for each $i = 1, 2, \ldots, m$. The kth-column, $k = 0, 1, \ldots, p$, of a pre-stacking is the sequence $(B_{1k}, B_{2k}, \ldots, B_{mk})$ where B_{ik} is the branch of G_i at a_{ik}. A <u>stacking</u> of $\{G_i \mid i = 1, 2, \ldots, m\}$ is a pre-stacking of the $\{G_i \mid i = 1, 2, \ldots, m\}$ with the branches of G in one-to-one correspondence with the columns of the stacking such that:

(1) If the branch of G corresponding to the kth-column of the stacking is B_k with t_k terminal edges of G in B_k then $m - t_k$ of the B_{ik}, $i = 1, 2, \ldots, m$, are isomorphic to B_k while the remaining t_k of the B_{ik} are in one-to-one correspondence with the $(B_k)_v$, v an endpoint of G in B_k such that corresponding ones are isomorphic (when we talk about isomorphisms of branches we always mean such that the roots correspond).

(2) For a fixed i, $B_{ik} = B_k$ for all but one k, $k = 0, 1, \ldots, p$.

(3) For any two i, j either:
(a) There are two integers s, t such that $B_{ik} = B_{jk}$ for $k \neq s, t$ and $B_{is} = B_s$, $B_{js} = (B_s)_v$ for some v in $V(B_s)$, $B_{it} = (B_t)_w$ for some w in $V(B_t)$, and $B_{jt} = B_t$.
Or
(b) There is one integer s such that $B_{ik} = B_{jk}$ for $k \neq s$ and $B_{js} = (B_s)_v$ for some v in $V(B_s)$, $B_{is} = (B_s)_w$ for some w in $V(B_s)$.

The <u>graph of a stacking</u> is the graph obtained by identifying the root of B_k to c_k in C_0. A <u>rotation</u> (<u>reflection</u>) of a G_i is a rotation (reflection), σ, of C_i such that $B_{ik} = B_{i,\sigma(k)}$ for all k.

<u>Lemma 4.11</u>. If H is a graph whose family of maximal subgraphs is $\{G_i \mid i = 1, 2, \ldots, m\}$, then H induces a stacking of the $\{G_i \mid i = 1, 2, \ldots, m\}$ whose graph is isomorphic to H.

<u>Proof</u>. Let the center of the pruned graph of H be the circuit $C_0 = (a_0, a_1, \ldots, a_p)$ such that the branch of H at a_k is B_k. Let t_k be the number of endpoints of H in B_k and let the endpoints be labeled so that the first t_0 are in B_0, the next t_1 are in B_1, etc., the last t_p are in B_p. Let the center of the pruned graph of G_i be $C_i = (a_{i0}, a_{i1}, \ldots, a_{ip})$ with the branch of G_i rooted at a_{ik} being labeled B_{ik} where $B_{ik} = (B_k)_{v_i}$ if v_i is in $V(B_k)$ and $B_{ik} = B_k$ if v_i is not in B_k. This is then

a stacking of $\{G_i \mid i = 1,2,\ldots,m\}$ whose graph is H. This stacking is said to be induced by H.

The remainder of the proof involves showing that the graph of any stacking of $\{G_i \mid i = 1,2,\ldots,m\}$ is isomorphic to G. It then follows from Lemma 4.11 that the family $\{G_i \mid i = 1,2,\ldots,m\}$ determines G up to isomorphism.

Let A be the stacking of $\{G_i \mid i = 1,2,\ldots,m\}$ induced by G as in the lemma. We now continue the proof of Theorem 4.10.

Let $G_1 = G_v$ and $G_2 = G_w$ (v and w as at the beginning of the proof). Suppose S is a stacking of $\{G_i \mid i = 1,2,\ldots,m\}$. We need to show that the graph of stacking S is isomorphic to the graph of stacking A. By the nature of the stacking we see that stacking S can be obtained from stacking A through a rotation or a reflection of each G_i, $i = 1, 2, \ldots, m$. Let σ be a reflection or rotation of G_1 in stacking A such that the branch at a_{1k} in G_1 of stacking A is isomorphic to the branch at $a'_{1,\sigma(k)}$ in G_1 of stacking S. Then by applying the inverse of σ to each G_i in stacking S we get a stacking whose graph is isomorphic to the graph of stacking S, except now we can assume that C_1 has the same labeling in this stacking as in stacking A. So without loss of generality we can assume that C_1 has the same labeling in each stacking. Let $(a_{10}, a_{11}, \ldots, a_{1p})$ be the labeling of C_1 in stacking A and in stacking S. Let $(a_{i0}, a_{i1}, \ldots, a_{ip})$ be the labeling of C_i in stacking A and let $(a'_{i0}, a'_{i1}, \ldots, a'_{ip})$ be the labeling of C_i in stacking S with B'_{ik} the branch of G_i at a'_{ik}. Note that stacking A has $B_{10} = B_v$ and $B_{20} = B_w$ since it is the induced stacking of Lemma 4.11.

We now consider two cases corresponding to the possibilities for labeling C_2 of stacking S with respect to the labeling of C_2 in stacking A.

Case 1. G_2 was reflected; say $a'_{20} = a_{2r}$ and hence $a'_{2k} = a_{2,r-k}$ $k = 0, 1, \ldots, p$, where the second subscripts are reduced mod $(p+1)$.

Subcase (1.1). Suppose there is an s and a t such that $B'_{2k} = B_{1k}$, $k \neq s, t$, $B_{1s} = B_v$, $B_{1t} = B$, $B'_{2s} = B$, and $B'_{2t} = B_w$.

Since $B \neq B_v$ we have $t \neq 0, s$. If $t = r$ we have $B_{1r} = B$ and since $t \neq 0$ we get $B_{2r} = B$. From stacking A we know $B_{10} = B_v$ and from the reflection we get $B'_{20} = B_{2r}$ and hence $B'_{20} \neq B_{10}$. But then we must have $s = 0$ since $B'_{2r} = B_{1r}$ for $k \neq s, t$. This implies that the graph of stacking A is isomorphic to the graph of stacking S. If $s = r$ we get $B'_{2r} = B'_{2s} = B$ and $B'_{2r} = B_{20} = B_w$ which is a contradiction. If $r = 0$ we get, $B_v = B_{10} = B'_{20} = B_{20} = B_w$, another contradiction. Thus we can assume that the integers $0, r, s$ and t are all distinct. But then we get: $B_v = B_{10} = B'_{20} = B_{2r} = B_{1r} = B'_{2r} = B_{20} = B_w$, which is a contradiction.

Subcase (1.2). Suppose there is an s such that $B'_{2k} = B_{1k}$ for $k \neq s$ and $B'_{2s} = B_w$, $B_{1s} = B_v$. If $s = 0$ we are finished. So assume $s \neq 0$ and hence $r \neq 0$. If $s \neq r$ we have: $B_w = B_{20} = B'_{2r} = B_{1r} = B_{2r} = B'_{20} = B_{10} = B_v$, which is a contradiction. Thus $s = r$ and we have $B_v = B_{10} = B'_{20} = B_{2r} = B_{1r} = B_{1s}$ and for $k \neq 0$, r we have $B_{1k} = B'_{2k} = B_{2,r-k} = B_{1,r-k}$. Hence the given reflection is an automorphism of G_1, such that $a_{10} \to a_{1r}$. Hence the graph of stacking S is isomorphic to the graph of stacking A.

Case 2. G_2 was rotated; say $a'_{2k} = a_{2,k+r}$ where the second subscripts are reduced mod $(p+1)$.

Subcase (2.1). Suppose there is an s and a t such that $B_{1,k} = B'_{2,k}$ for $k \neq s, t$ and $B_{1,s} = B_v$, $B_{1,t} = B$, $B'_{2,s} = B$ and $B'_{2,t} = B_w$. We need the following lemma:

Lemma 4.12. Let $R = \langle r \rangle = \{x$ in $Z_{p+1}: x \equiv tr \bmod (p+1)$, t an element of $Z\}$. We have
    (1) $B_{1,mr} = B$, B_v, or B_w for all m
    (2) If x is not an element of R then $B_{1,x} = B_{1,y}$ for all
        y in $R + x$.

Proof. We have the following sequence:
$B_v = B_{10}$, $B'_{20} = B_{2r}$, B_{1r}, $B'_{2r} = B_{2,2r}$, $B_{1,2r}$, $B_{1,2r}$, $B'_{2,2r} = B_{2,3r} \cdots, B_{1,kr}$, $B'_{2,kr} = B_{2,(k+1)r}$, $B_{1,(k+1)r} \cdots, B_{2,t_0 r} = B_{2,0} = B_w$ where t_0 is the order of r, $B_{1,k} = B'_{2,k}$ for $k \neq s, t$ and $B_{2,k} = B_{1,k}$ for $k \neq 0$. Suppose for some $m > 0$ we have $B_{1,mr} \neq B$, B_v, or B_w. Then $mr \bmod (p+1) \neq 0$, s, or t. Therefore we have $B_{1,mr} = B'_{2,mr} = B_{2,(m+1)r}$ and so $B_{2,(m+1)r} \neq B$, B_v, or B which implies $(m+1)r \bmod (p+1) \neq 0$, s, or t. Therefore $B_{2,(m+1)r} = B_{1,(m+1)r}$. As a consequence $B_{1,(m+1)r} \neq B$, B_v, or B_w. But this is impossible since for some k we have $m + k \equiv t_0$ and hence $B_{1,(m+k)r} = B_{20} = B_w$. If x is not an element of R, then $x + yr$ is not an element of R for any y. Hence using the relation between stacking A and stacking S we have: $B_{1x} = B_{1,x+r} = B_{1,x+2r} = \cdots$ as asserted. To illustrate the symmetry of G_1 we add the following without proof:

There are integers m and n such that;
$$B_v = B_{10} = B_{1r} = B_{1,2r} = \cdots = B_{1,mr}$$
$$B = B_{1,(m+1)r} = \cdots = B_{1,nr} \quad \text{and if} \quad n < t_0 - 1$$
$$B_w = B_{1,(n+1)r} = \cdots = B_{1,(t_0-1)r} \quad \text{where} \quad s \equiv mr \bmod (p+1),$$
$t \equiv nr \bmod (p+1)$ and t_0 is the order of r.

We continue now with the proof of Subcase (2.1). As in Subcase (1.1) we see that $t \neq 0$ and if $s = 0$ we are finished. So suppose $t \neq 0$ and $s \neq 0$. Thus by the definition of a stacking we have: $B_{10} = B'_{20} = B_v$, and $B'_0 = B_v$. Since B_v has at least one more endpoint, namely w, we have for some i, $3 \leq i \leq m$, $B'_{i0} = (B_v)_w$. Note that, since $B'_0 = B_v$ and $B'_k = B'_{ik}$ for $k \neq 0$, the graph of stacking S is isomorphic to the graph obtained from G_i by replacing the branch B'_{i0} with a B_v type branch. We will show that the graph obtained from G_i in this way is isomorphic to G.

Since G_i has one more $(B_v)_w$ type branch than G and G_1 has one more B_v type branch than G we see by the definition of stacking A that there is an n and a q such that $B_{1k} = B_{ik}$ for $k \neq n$, q, $B_{1n} = B_v$, $B_{in} = (B_v)_w$, $B_{1q} = B_v$, $B_{iq} = B$. And G is obtained from G_i by replacing the branch B_{in} with a B_v type branch. But from stacking S we also know the following:

(1) If $B_{1k} = (B_v)_w$ then $B_{1x} = (B_v)_w$ for all x in $k + R$

(2) If $B'_{ik} = (B_v)_w$ for k not in R then $B'_{ix} = (B_v)_w$ for all x in $k + R$

(3) $B'_{i0} = (B_v)_w$ but $B'_{ix} \neq (B_v)_w$ for x in R, $x \neq 0$.

Thus we must have $B'_{i0} = B_{in}$ since $|R| \geq 3$. Hence the graph of stacking S is the same as that of stacking A.

<u>Subcase (2.2)</u>. Suppose there is an s such that $B'_{2k} = B_{1k}$ for $k \neq s$ and $B'_{2s} = B_w$, $B_{1s} = B_v$.

The proof of Subcase (2.2) is very similar to the proof of Subcase (2.1) and so it is omitted.

That completes the proof of Theorem 4.10.

<u>Definition</u>. If G is a graph on n vertices and $\Gamma(G)$ is a subgroup of the dihedral group on n points then we say that G is quasi-dihedral.

<u>Corollary 4.13</u>. If G has at least three branches with tree growths and if the pruned center of G is quasi-dihedral, then G can be reconstructed from $\{G_v \mid v \in S(G)\}$.

<u>Proof</u>. The proof is that of Theorem 4.10 since the only use we made of the fact that the pruned center was a circuit was that an automorphism of it was either a rotation or reflection.

Definition. A cactus is a connected graph G such that each block of G is either an edge or a circuit.

Corollary 4.14. The vertex problem is true for cacti and if G is a cactus with a terminal vertex then the terminal-vertex problem is true for G, provided G is not one of the counterexamples described in Theorems 4.1, 4.2, 4.3, 4.4.

Proof. The latter assertion follows immediately from the results of this section while the former follows from the latter and Theorem 3.5. One needs to verify, of course, that the counterexamples to the terminal vertex problem are not counterexamples to the vertex problem.

The first assertion of the corollary has been proved independently by Geller and Manvel in [3].

Corollary 4.15. The edge problem is true for cacti.

Proof. This follows from Corollary 4.14 unless the cactus G has no terminal vertices. But then there exist a G^e, $e \in E(G)$, that has two terminal vertices. We reconstruct G by adding an edge between these two vertices. Again, the counterexamples to the terminal vertex problem must be checked.

5. Directed graphs. Although most of the work on the reconstruction problem has been for undirected graphs there have been some results for directed graphs.

An oriented graph is obtained from a graph when each edge is assigned a unique direction. A signed graph has the numbers + 1 or -1 assigned to each of its edges. A tournament is an oriented complete graph. A directed graph G is strong if for each $u, v \in V(G)$ there is a directed path from u to v.

In [4], Harary and Palmer point out that by the same proof as that of Theorem 3.8 one gets:

Corollary 5.1. If T is an oriented (signed) tree with at least three endpoints, then T is determined by the maximal oriented (signed) subgraphs of T.

Similarly, from Theorem 4.10 we get:

Corollary 5.2. If G is an oriented (signed) graph such that at least three branches of G have tree growths and if the pruned center of G is quasi-dihedral, then G can be reconstructed from the maximal oriented (signed) subgraphs of G.

For tournaments we have:

Theorem 5.3 (Harary and Palmer [6]). The edge problem is true for tournaments.

Theorem 5.4 (Harary and Palmer [6]). If T is a tournament with $|V(T)| \geq 5$ and if T is not strong, then the vertex problem is true for T.

In [4], Harary and Palmer give a strong tournament on five vertices for which the vertex problem is not true. E.T. Parker has recently constructed another such tournament on seven vertices. However these examples do not suggest any class of counterexamples, so it is not clear whether they are just isolated abberations or not. At any rate, the vertex problem for strong tournaments seems to offer many hours of frustrating combat for the willing challenger.

6. <u>When is a given collection the maximal subgraphs of some graph?</u>
There is one aspect of the reconstruction problem which has received only scant mention: namely, if $\{G_i \mid i = 1,2,\ldots,m\}$ is a given collection of graphs does there exist a graph G with $V(G) = \{v_i: i = 1, 2, \ldots, m\}$ and $G_{v_i} = G_i$, $i = 1, 2, \ldots, m$. There is of course a corresponding edge problem and as before we could require that the G_i are obtained by deleting only vertices of degree one from G, etc. We call G a <u>predecessor</u> of $\{G_i \mid i = 1,2,\ldots,m\}$ with respect to the appropriate type of vertices.

Most of the previous results, as typified by Kelly's Lemma, are predicated on the existence of at least one such predecessor.

In this section we indicate how one uses the concept of stackings to characterize collections $\{G_i \mid i = 1,2,\ldots,m\}$ having a connected predecessor with terminal vertices and circuits.

If G is a connected graph with n vertices and k edges, with circuits, with terminal vertices $\{v_i: i = 1, 2, \ldots, m\}$, and with a block C as its pruned center then for each $i = 1, 2, \ldots, m$, G_{v_i} is a connected graph with $n - 1$ vertices and $m - 1$ edges, with circuits, and with a block $C_i = C$ as its pruned center.

<u>Theorem 6.1.</u> Let $\{G_i \mid i = 1,2,\ldots,m\}$ be a collection of graphs such that they all have the same number of vertices and the same number of edges, each has circuits, and each has a block C_i as its pruned center. Then there is a graph G with terminal vertices $\{v_i: i = 1,2,\ldots,m\}$ such that $G_{v_i} = G_i$ for $1 \le i \le m$ if and only if there exists a labeling of the vertices of C_i as $\{a_{ik}: 1 \le k \le p\}$ for $1 \le k \le p$; and isomorphisms $\sigma_i: C_1 = C_i$ with $\sigma_i(a_{1k}) = a_{ik}$ such that if B_{ik} is the branch of G_i rooted at a_{ik} then for $i \notin T_k$ the terminal vertices of B_{ik} (other than a_{ik}) can be labeled as $\{v_j: 1 \le j \le |T_k|\}$ so that $(B_{ik})_{v_j} = B_{jk}$ for $j \in T_k$ while $B_{ik} = B_{jk}$ for $j \notin T_k$.

<u>Proof.</u> The conditions given are an internal description of a stacking of $\{G_i \mid i = 1,2,\ldots,m\}$ and as in Section 4 we see that a graph G as above induces such a stacking and conversely the graph of such a stacking A induces stacking A.

It is clear that one can, in the same manner, give a characterization of the families $\{G_i \mid i = 1,2,\ldots,m\}$ having a predecessor (with respect to the terminal vertices) that is a connected graph with circuits, terminal vertices, and that has a cut-point as its pruned center.

Thus if a collection $\{G_i \mid i = 1,2,\ldots,n\}$ (with $\{G_i \mid i = 1,2, \ldots,m\}$ being those with the maximal number of edges) has a predecessor, with respect to the set of all vertices, that is a connected graph with circuits and terminal vertices then the collection $\{G_i \mid i = 1,2,\ldots,m\}$ has the same predecessor, with respect to the terminal vertices. So a solution to the problem of characterizing the collections having a predecessor, with respect to all vertices, can be given in terms of the edge maximal members of the collection.

Again making use of the fact that $G^e = G_a \cup \{a\}$ if $\deg(a) = 1$ and $e = (a,b) \in E(G)$ we can solve the corresponding edge problem.

REFERENCES

1. J.A. Bondy, On Kelly's congruence theorem for trees, <u>Proc. Cambridge Phil. Soc</u>. (to appear).

2. J.A. Bondy, On Ulam's conjecture for separable graphs (to appear).

3. D. Geller and B. Manvel, Reconstruction of cacti, <u>Canad. J. Math</u>. (to appear).

4. F. Harary and E. Palmer, The reconstruction of a tree from its maximal subtrees, <u>Canad. J. Math</u>., 18 (1966), 803-811.

5. F. Harary and E. Palmer, On similar points of a graph, <u>J. Math. Mech</u>., 15 (1966), 623-630.

6. F. Harary and E. Palmer, On the problem of reconstructing a tournament from subtournaments, <u>Monatsh. für Math</u>., 71 (1967), 14-23.

7. F. Harary, On the reconstruction of a graph from a collection of subgraphs, in <u>Theory of graphs and its applications</u> (M. Fielder, ed.) Prague, 1964, 47-52.

8. R.L. Hemminger, On reconstructing a graph, <u>Proc. Amer. Math. Soc</u>., 20, (1969).

9. P.J. Kelly, A congruence theorem for trees, <u>Pacific J. Math</u>., 7 (1957), 961-968.

10. B. Manvel, Reconstruction of trees (to appear).

11. O. Ore, <u>Theory of graphs</u>, Colloq. Pub., 38, Amer. Math. Soc., Providence, R.I., 1962.

12. S.M. Ulam, <u>A collection of mathematical problems,</u> New York, 1960, p. 29.

13. H. Whitney, Congruent graphs and the connectivity of graphs, Amer. J. Math., 54 (1932), 150-168.

INCIDENCE PATTERNS OF GRAPHS AND COMPLEXES[1]

Branko Grünbaum, University of Washington

1. **Introduction.** One of the principal reasons for the applicability of graph theory to other mathematical and non-mathematical disciplines stems from the possibility of representing by graphs certain significant relational patterns of the objects under investigation. It is not surprising, therefore, that the formation of patterns from given collections of objects has become a frequent topic in graph theory, applied to graphs as well as to other objects.

One of the aims of the present paper is to give a survey of results known on some of the patterns that have been investigated.

To be precise, by an <u>incidence pattern</u> P I mean the association, according to definite rules which depend on the pattern in question, of a graph or complex P(C) with any given graph, complex, or similar object C of a certain type. The idea is to find in the pattern a reflection of some of the incidence properties of the original graph or complex. To fix the ideas, let me recall the simplest example of such a pattern - the formation of the interchange graph of a given graph.

The interchange graph (or line-graph) I(G) of a graph G has vertices which correspond to the edges of G, two vertices of I(G) determining an edge of I(G) if and only if the corresponding edges of G have a common vertex.

The literature on interchange graphs, and on other incidence patterns, is almost unbelievably large. It is my hope that the juxtaposition of the various patterns and of the results known about them will also serve as a guide to the practically inexhaustible supply of open problems, and help to relate the various patterns among themselves and to other fields.

It was found convenient to divide the survey into two parts. The first part is concerned with interchange graphs and some of their generalizations and analogs, while the second deals mainly with nerves and their 1-skeleta.

2. **Interchange graphs and related notions.** The basic results on interchange graphs may be formulated as follows:

Theorem 1. A graph H is (isomorphic to) the interchange graph I(G) of some graph G if and only if there exists an edge-disjoint collection of complete graphs covering H, such that each vertex of H belongs to at most two members of the collection.

[1]Research supported in part by Office of Naval Research contract N00014-67-A-0103-0003.

Theorem 2. A graph H is the interchange graph of some graph if and only if none of the spanned subgraphs of H coincides with any of the graphs in Figure 1.

Theorem 3. If the interchange graph I(G) is connected, then it determines the graph G uniquely except when I(G) is the graph of Figure 2a, in which case G may be either the graph of Figure 2a or that of Figure 2b.

For the case of interchange graphs of graphs, Theorems 1 and 2 provide answers (in different terms) to the first of the following general problems, while Theorem 3 solves the second. Formulated for an incidence pattern P(C) of complexes C, the general questions are:

Characterization problem. For a given pattern P, characterize those complexes K which are (isomorphic to) P(C) for some complex C (belonging to a certain family of complexes).

Determination problem. For a given pattern P, and for a complex K such that K = P(C), to what extent is C characterized by K ? In other words, what is the relation between complexes which have isomorphic patterns?

These two questions have been investigated for many incidence patterns. However, before proceeding to an account of these investigations, we will briefly mention the results on interchange graphs we found in the literature.

Theorems 1, 2, and 3, or some of them, and other characterizations of interchange graphs of graph, may be found in Andreatta [1], Beineke [1,2], Berge [1], Busacker - Saaty [1], Chartrand [1], Harary [3], Heuchenne [1], Krausz [1], Ore [1], Ray-Chaudhuri [1], Rooij - Wilf [1], Sabidussi [1], Seshu - Reed [1], Whitney [1].

Euler and Hamilton circuits in interchange graphs and in repeated interchange graphs were considered by Chartrand [2,3], Harary [3], Harary - Nash-Williams [1], Sedlacek [1].

Graphs isomorphic with their interchange graphs were discussed by Ghirlanda [1] and Menon [1,2].

Sedlacek [1] (see also Ore [2]) characterized planar interchange graphs. Various other properties of interchange graphs were considered by Behzad [1], Behzad - Chartrand - Nordhaus [1], Chartrand - Stewart [2] and Kotzig [1].

Suitably defined interchange graphs of directed graphs were considered by Beineke [2], Chartrand - Stewart [1], Geller - Harary [1], Harary - Norman [1], Kasteleyn [1], and Muracchini - Ghirlanda [1].

It is possible to generalize in many ways the method of formation of interchange graphs of graphs. One of the simplest generalizations is the following:

Let C be a simplicial complex, and let k be a positive integer. The kth interchange graph $I_k(C)$ of C has vertices which

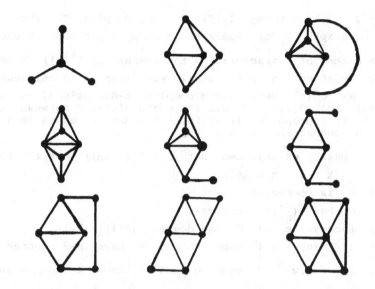

Figure 1

(a)   (b)

Figure 2

are in a biunique correspondence with the k-simplices of C,   two
vertices of $I_k(C)$ being connected by an edge if and only if the
corresponding k-simplices of C have a common (k-1)-simplex.

Clearly, if C is a graph then $I_1(C) = I(C)$.

The trivial part of Theorem 1 generalizes as follows:
(*)   Each $I_k(C)$ may be covered by a family of edge-disjoint complete graphs such that each vertex of $I_k(C)$ belongs to at most
k+1 members of the family.

However, for k > 1 the condition (*) is far from sufficient.
For example, the graph of Figure 3 is easily seen to satisfy (*)

for k = 2 without being $I_2(C)$ for any complex C. The characterization of graphs $I_k(C)$ seems to be an open problem for each k ≥ 2.

One method of characterizing the graphs $I_k(C^k(n))$ received
considerable attention, particularly with respect to uniqueness.
(Recall that $C^k(n)$ denotes the complete k-dimensional complex on n
vertices.) Let d(u,v) denote the graph-distance between the vertices u,v in a graph G (that is, the number of edges in a shortest
path in G connecting u and v).

It is easily established that $G = I_k(C^k(n))$ has the following
properties (k ≥ 1, n ≥ k+1) :

    (i)   G is connected.

    (ii)   G has $\binom{k}{n+1}$ vertices.

    (iii)   Each vertex of G has degree (k+1)(n-k-1).

    (iv)   If d(u,v) = 1 then u and v have n-2 common neigh-
bors.

    (v)   If d(u,v) = 2 then u and v have 4 common neighbors.

The above properties do not fully characterize interchange
graphs of complete complexes. The known results are:

<u>Theorem 4</u>. If G is a graph which, for some integers k and n,
satisfies conditions (i) - (v), then G is uniquely determined and
is isomorphic to $I_k(C^k(n))$ provided either

                (a)   n > 2k(k+1) + 4,

or   (b)   k = 2 and n ≤ 8,

or   (c)   k = 1 and n ≤ 7 .

<u>Theorem 5</u>. For k = 1 and n = 8 there exist graphs G which have
properties (i) - (v) but are not isomorphic to $I(C^1(8))$.

Figure 3

Various parts of these results were established by Aigner [1,2],
Bose - Laskar [1], Chang [1,2], Connor [1], Dowling [1], Hoffman
[1,2], and Shrikhande [1].

The situation is similar for interchange graphs of complete bipartite graphs $B(n,m) = C^0(n) \vee C^0(m)$, there being only one exceptional case (n,m) = (4,4) (see Hoffman [3], Moon [1], Shrikhande

[2]). Certain other families of graphs seem to behave analogously (Aigner [3], Laskar [1]). Interchange graphs of higher-dimensional complete multipartite complexes do not seem to have been investigated.

In a spirit similar to that of Theorems 4 and 5 the incidence properties of finite affine and projective planes have been consider-ed (Hoffman [4], Hoffman - Ray-Chaudhuri [1]). There is also a siz-able literature on an analogous treatment of balanced incomplete block designs (see, for example, Bose [1], Das [1], Hoffman - Ray-Chaudhuri [2]).

The scope of the definition of $I_k(C)$ may be widened by per-mitting C to be any cell complex. The inherent interest in this generalization is evident from the fact that if C is a cell-decom-position of any 2-manifold then $I_2(C)$ is the dual graph (in that manifold) of the graph (i.e., 1-skeleton) of C. Rather trivially we also have

Theorem 6. For every $k \geq 2$, each graph G is isomorphic to $I_k(C)$ for a suitable k-dimensional cell complex C.

Indeed, if $k = 2$ we may take, for each vertex v of G of degree d, a 2d-sided polygon, and identify the d even-numbered edges of the polygon in pairs with the appropriate edges of the poly-gons constructed for the vertices to which v is adjacent (a trivial modification takes care of the case $d = 1$). For $k > 2$ the poly-gons may obviously be "thickened" in a suitable way.

The complexes just constructed are topological (or piecewise-linear) cell complexes. An interesting (but seemingly very hard) problem is the characterization of the interchange graphs of geomet-ric cell complexes.

The notion of interchange graphs of cell complexes becomes much more interesting if the complexes under consideration are suitably restricted. One such possibility is to start with a (k+1)-dimension-al convex polytope P and its boundary complex $B(P)$. Then it is easy to see that $I_k(B(P))$ is just the 1-skeleton of the (k+1)-polytope P^* dual to P. Hence the questions of determining prop-erties or characterizations of interchange graphs of the boundary complexes of polytopes are equivalent to the corresponding questions about the graphs (i.e., 1-skeletons) of polytopes dual to the origi-nal ones. For example, the case of 3-polytopes is settled by the following reformulation of the famous theorem of Steinitz (see Steinitz [1], Steinitz - Rademacher [1], Grünbaum [2], Barnette - Grünbaum [1]):

Theorem 7. A graph G is isomorphic to $I_2(B(P))$ for some 3-poly-tope P if and only if G is planar and 3-connected. Moreover, G determines the combinatorial type of P.

The graphs of higher-dimensional polytopes, and thus also the interchange graphs of their boundary complexes, have not been com-pletely characterized. References to the voluminous literature deal-ing with properties of such graphs may be found in Grünbaum [2,4].

Two analogs of interchange graphs have been used in the investigation of planar graphs.

If G is a planar graph, the medial graph M(G) of G (see Ore [2]) has vertices which correspond to the edges of G, two vertices of M(G) being connected if and only if the corresponding edges of G have a common endpoint and are on the boundary of a common face (country). In connection with a proof of Theorem 7, medial graphs were thoroughly investigated by Steinitz [1] (see also Steinitz [1], Grünbaum [2]).

The radial graph R(G) of a planar graph G may be defined as the dual (in the sense of planar graphs) of the graph M(G). As is easily seen (Ore [2]), we have

Theorem 8. A planar graph H is R(G) for some 2-connected planar graph G if and only if H is a maximal, bipartite, planar graph.

Radial graphs may be used to simplify proofs of Steinitz's Theorem 7 (see Barnette - Grünbaum [1]).

Another incidence pattern somewhat similar to interchange graphs is that of the total graphs T(G) of a graph G. The vertices of T(G) correspond to all the faces (vertices and edges) of G, two vertices of T(G) determining an edge if and only if the two corresponding faces of G are either incident or adjacent. Though introduced only recently (Behzad [1]), there is already a rather voluminour literature devoted to them (Behzad [1,2], Behzad - Chartrand [1,2], Behzad - Chartrand - Cooper [1], Behzad - Chartrand - Nordhaus [1], Chartrand - Stewart [1], Gupta [1], Rosenfeld [1], Zykov [1]). The results obtained parallel to a large extent those on interchange graphs of graphs; unfortunately, no satisfactory analog of Theorems 1 and 2 is known for total graphs.

3. Nerves and related patterns. The formation of the interchange graph I(G) of a graph G may be interpreted as follows: We are given a family F of objects (the edges of G) and we assign to each of them a vertex; two of those vertices determine an edge if and only if the corresponding objects in F have a non-empty intersection. For an arbitrary family F of objects, we may take the above sentence as the definition of a new graph J(F) which, for lack of a better name, we shall call the intersection graph of the family F (see Harary [3]).

Clearly, the wealth of problems one may consider is staggering; relatively few have been considered in any detail.

The question of finding "small" families F which have a given graph G as their intersection graph was considered by Erdös - Goodman - Pósa [1] and Harary [3].

Hamelink [1] investigated "clique-graphs", that is graphs J(F) where F is the family of cliques (i.e., maximal complete subgraphs) of a graph G.

The "block graph" of a graph G is the graph J(F) where F is the family of blocks of G; block graphs were characterized and discussed by Harary [1,2,3].

The <u>nerve</u> N(C) of C is defined as the nerve N(F) of that family F. (See Figure 4 for an illustration.)

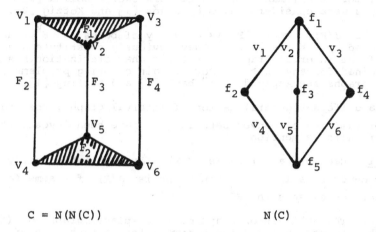

$$C = N(N(C)) \qquad\qquad N(C)$$

Figure 4

The following results are among those established in Grünbaum [3].

<u>Theorem 3</u>. A simplicial complex K is isomorphic with the nerve N(C) of some n-dimensional simplicial complex C if and only if K satisfies the conditions
    (i)   Each edge of K belongs to at most n facets of K.
    (ii) Each vertex of K belongs to at most n+1 facets of K.

Let a simplicial complex C be called <u>taut</u> provided each vertex v of C is the intersection of all the facets of C that contain v. Note that triangulated manifolds or pseudomanifolds are taut, as are many other interesting types of complexes (see Figure 4).

<u>Theorem 4</u>. If C is a taut complex, then N(C) is taut, and C is isomorphic to N(N(C)). Conversely, if C is isomorphic to N(N(C)), then C is taut.

In other words, the nerve-operation defines a duality between taut complexes. Moreover, calling <u>strict face</u> of C any face of C that is the intersection of facets of C, we have:

<u>Theorem 5</u>. The strict faces of each taut complex C are in a one-to-one, inclusion-reversing correspondence with the strict faces of N(C).

The characterization of all complexes with a given nerve is easily accomplished, as is the proof of the fact that C and N(C) are homotopic.

We conclude by observing that the nerve operation has some

Motivated by a problem in genetics (see Benzer [1,2]), characterizations of "interval graphs", that is graphs $J(F)$ for families
F of intervals on the real line, were obtained by Fulkerson - Gross
[1,2], Gilmore - Hoffman [1], and Lekkerkerker - Boland [1]. The
same graphs were considered also by Hajós [1] and Kotzig [2].

Graphs $J(F)$, where F is a family of boxes in E^k, have been
investigated by Roberts [1,2]. They obviously constitute a generalization of the interval graphs, but their characterization has not
been obtained for any $k \geq 2$. For certain coloring problems concerning such graphs see Asplund - Grünbaum [1] and Burling [1].

As a different generalization of interval graphs, the graphs
$J(F)$ for families of convex sets in E^d have been investigated. As
typical results we mention:

Theorem 1. Each planar graph is $J(F)$ for a suitable family F of
compact convex sets in E^2; each graph is $J(F)$ for some family F
of compact convex sets in E^3.

Theorem 2. The subdivision graph of a non-planar graph G (that is,
the graph obtained by inserting a new vertex in each edge of G) is
not of the form $J(F)$, for any family F of convex sets in the
plane.

For these results and various generalizations and analogous
questions see Wegner [1]; references to much of the older literature
may be found in Danzer - Grünbaum - Klee [1].

The graphs $J(F)$ for families of translates of one convex set
are of interest in connection with various problems in Minkowski
spaces, functional analysis, and combinatorial geometry. The reader
is referred to Danzer - Grünbaum - Klee [1] and Grünbaum [1] for details and references. For a novel application see Melzak [1].

The graph $J(F)$ may be considered as the 1-skeleton of a simplicial complex $N(F)$, the nerve of F, associated with each family
F. The vertices of $N(F)$ correspond to the members of the family
F, and a set of vertices of $N(F)$ determines a simplex of $N(F)$ if
and only if the corresponding members of F have a non-empty intersection.

Nerves of families which are (open, or closed) covers of topological spaces have been extensively studied; almost all standard
texts on topology will lead the reader to the appropriate references.

Many of the papers mentioned above deal not only with $J(F)$ but
also with $N(F)$ for various types of families F of convex sets; in
particular, Wegner [1] contains numerous results in this direction.

Instead of detailing these results we shall turn to a specializ-
ed application of the nerve construction, which seems to be attrac-
tive from various points of view.

For a simplicial complex C, we consider the family F of
facets of C, that is the family of all maximal closed faces of C.

"stabilizing" features. More precisely,

Theorem 6. For any simplicial complex C the sequence C, N(C),
N(N(C)), ... yields, after finitely many steps, a taut complex homotopic to C.

REFERENCES

M. Aigner
    1. On the tetrahedral graph. (to appear)
    2. A characterization problem in graph theory. (to appear.)
    3. A characterization of a class of regular graphs. Abstract
        655-53. Notices Amer. Math. Soc. 15 (1968), 481.

A. Andreatta
    1. On finite graphs which are line-graphs. (Italian). Ist.
        Lombardo Accad. Sci. Lett. Rend. A. 98 (1964), 133-156.

E. Asplund and B. Grünbaum
    1. On a coloring problem. Math. Scand. 8 (1960), 181-188.

D.W. Barnette and B. Grünbaum
    1. On Steinitz's theorem concerning convex 3-polytopes and on
        some properties of planar graphs. (to appear)

M. Behzad
    1. Graphs and their chromatic numbers. Ph.D. Thesis, Michigan
        State University (1965).
    2. A criterion for the planarity of the total graph of a graph.
        Proc. Camb. Phil. Soc. 63 (1967), 679-681.

M. Behzad and G. Chartrand
    1. Total graphs and traversability. Proc. Edinburgh Math. Soc.
        15 (1966), 117-120.
    2. An introduction to total graphs, Théorie des Graphes, Rome
        1966, 31-33.

M. Behzad, G. Chartrand and J.K. Cooper, Jr.
    1. The colour numbers of complete graphs. J. London Math. Soc.
        42 (1967), 226-228.

M. Behzad, G. Chartrand and E.A. Nordhaus
    1. Triangles in line-graphs and total graphs. Indian J. Math.
        (to appear)

L.W. Beineke
    1. A characterization of derived graphs. (to appear)
    2. On derived graphs and digraphs. Beiträge zur Graphentheorie.
        (Edited by H. Sachs, H.-J. Voss, and H. Walther) Teubner,
        Leipzig 1968. pp. 17-23.

S. Benzer
    1. On the topology of the genetic fine structure. Proc. Nat.
        Acad. Sci. U.S.A. 45 (1959), 1607-1620.
    2. The fine structure of the gene. Scientific American 206
        (1962), No. 1, 70-84.

C. Berge
    1. Färbungen von Graphen, deren sämtliche bzw. deren ungerade
       Kreise starr sind. Wiss. Zeitschr. Martin Luther Univ.
       Halle-Wittenberg, 10 (1961), 114-115.

R.C. Bose
    1. Strongly regular graphs, partial geometries and partially
       balanced designs. Pacif. J. Math. 13 (1963), 389-419.

R.C. Bose and R. Laskar
    1. A characterization of tetrahedral graphs. J. Combinatorial
       Theory 3 (1967), 366-385.

J.P. Burling
    1. On coloring problems of families of prototypes. Ph.D. Thesis,
       University of Colorado (1965).

R.G. Busacker and T.L. Saaty
    1. Finite graphs and networks. McGraw-Hill, New York, 1965.

L.-C. Chang
    1. The uniqueness and non-uniqueness of the triangular associ-
       ation schemes. Science Record, Math. N.S. 3 (1959), 604-613.
    2. Association schemes of partially balanced designs with para-
       meters $v = 28$, $n_1 = 12$, $n_2 = 15$ and $p_{11}^2 = 4$. Science Rec-
       ord, Math., N.S. 4 (1960), 12-18.

G. Chartrand
    1. Graphs and their associated line-graphs. Ph.D. Thesis,
       Michigan State University (1964).
    2. The existence of complete cycles in repeated line-graphs.
       Bull. Amer. Math. Soc. 71 (1965), 668-670.
    3. On hamiltonian line-graphs. Trans. Amer. Math. Soc. 134
       (1968), 559-566.

G. Chartrand and M.J. Stewart
    1. Total digraphs. Canad. Math. Bull. 9 (1966), 171-176.
    2. The connectivity of line-graphs. (to appear)

W.S. Connor
    1. The uniqueness of the triangular association scheme. Ann.
       Math. Stat. 29 (1958), 262-266.

L. Danzer, B. Grünbaum and V. Klee
    1. Helly's theorem and its relatives. Proc. Symp. Pure Math.
       vol. 7 (Convexity), pp. 101-180 (1963).

B. Das
    1. Tactical configurations and graph theory. Calcutta Statist.
       Assoc. Bull. 16 (1967), 136-138.

T.A. Dowling
    1. A characterization of the graph of the T_m association
       scheme. Notices Amer. Math. Soc. 14 (1967), 843.

P. Erdös, A. Goodman and L. Pósa
    1. The representation of a graph by set intersections. <u>Canad.</u>
       <u>J. Math.</u>, 18 (1966), 106-112.

D.R. Fulkerson and O.A. Gross
    1. Incidence matrices with the consecutive ones property. <u>Bull.</u>
       <u>Amer. Math. Soc.</u> 70 (1964), 681-684.
    2. Incidence matrices and interval graphs. <u>Pacif. J. Math.</u> 15
       (1965), 835-855.

D. Geller and F. Harary
    1. Arrow diagrams are line digraphs. <u>J. SIAM Appl. Math.</u> 16
       (1968), 1141-1145.

A. Ghirlanda
    1. Osservazioni sulle caratteristiche dei grafi o singrammi.
       <u>Ann. Univ. Ferrara Sez.</u> VII (N.S.) 11 (1962/65), 93-106.

P.C. Gilmore and A.J. Hoffman
    1. A characterization of comparability graphs and of interval
       graphs. <u>Canad. J. Math.</u> 16 (1964), 539-548.

B. Grünbaum
    1. Borsuk's problem and related questions. <u>Proc. Symp. Pure</u>
       <u>Math.</u> vol. 7 (Convexity), pp. 271-284 (1963).
    2. <u>Convex polytopes.</u> Wiley, New York, 1967.
    3. Nerves of simplicial complexes. <u>Aequationes Math.</u> (to ap-
       pear)
    4. Polytopes, graphs, and complexes. <u>Bull. Amer. Math. Soc.</u>
       (to appear).

R.P. Gupta
    1. The cover index of a graph. (to appear)

G. Hajós
    1. On a type of graph. <u>Inter. Math. News. Sondernummer</u> (1957),
       65.

R.C. Hamelink
    1. A partial characterization of clique graphs. <u>J. Combinator-</u>
       <u>ial Theory</u> 5 (1968), 192-197.

F. Harary
    1. An elementary theorem on graphs. <u>Amer. Math. Monthly</u> 66
       (1959), 405-407.
    2. A characterization of block-graphs. <u>Canad. Math. Bull.</u> 6
       (1963), 1-6.
    3. <u>Graph Theory.</u> Addison-Wesley, Reading (to appear).

F. Harary and C. Nash-Williams
    1. On eulerian and hamiltonian graphs and line graphs. <u>Canad.</u>
       <u>Math. Bull.</u> 8 (1965), 701-709.

F. Harary and R. Norman
    1. Some properties of line digraphs. <u>Rend. Circ. Mat. Palermo</u>
       9 (1961), 161-168.

C. Heuchenne
1. Sur une certaine correspondence entre graphes. Bull. Soc. Roy. Sci. Liege 33 (1964), 743-753.

A.J. Hoffman
1. On the exceptional case in a characterization of the arcs of a complete graph. IBM Journal 4 (1960), 487-496.
2. On the uniqueness of the triangular association scheme. Ann. Math. Stat. 31 (1960), 492-497.
3. On the line-graph of the complete bipartite graph. Ann. Math. Stat. 35 (1964), 883-885.
4. On the line-graph of a projective plane. Proc. Amer. Math. Soc. 16 (1965), 297-302.

A.J. Hoffman and D. Ray-Chaudhuri
1. On the line-graph of a finite affine plane. Canad. J. Math. 17 (1965), 687-694.
2. On the line-graph of a symmetric balanced incomplete block design. Trans. Amer. Math. Soc. 116 (1965), 238-252.

P. Kasteleyn
1. A soluble self avoiding walk problem. Physica 29 (1963), 1329-1337.

A. Kotzig
1. On the theory of finite regular graphs of degree three and four. (Slovak) Časop. Pěst. Mat. 82 (1957), 76-92.
2. Paare Hajóssche Graphen. Časop. Pěst. Mat. 88 (1963), 236-241.

J. Krausz
1. Démonstration nouvelle d'un théorème de Whitney sur les réseaux. (Hungarian; French summary) Mat. Fiz. Lapok 50 (1943), 75-85.

R. Laskar
1. A characterization of cubic lattice graphs. J. Combinatorial Theory 3 (1967), 386-401.

C.G. Lekkerkerker and J.C. Boland
1. Representation of a finite graph by a set of intervals on the real line. Fund. Math. 51 (1962), 45-64.

Z.A. Melzak
1. On a class of configuration and coincidence problems. Bell System Techn. J. 47 (1968), 1105-1129.

V. Menon
1. The isomorphism between graphs and their adjoint graphs. Canad. Math. Bull. 8 (1965), 7-15.
2. On repeated interchange graphs. Amer. Math. Monthly 73 (1966), 986-989.

J.W. Moon
1. On the line-graph of the complete bigraph. Ann. Math. Stat. 34 (1963), 664-667.

L. Muracchini and A.M. Ghirlanda
1. Sul grafo commutato e sul grafo apposto di un grafo orient-
   ato. Atti Sem. Mat. Fis. Univ. Modena, 14 (1865), 87-97.

O. Ore
1. Theory of Graphs. Amer. Math. Soc. Colloq. Publ. vol. 38.
   Providence 1962.
2. The four-color problem. Academic Press, New York 1967.

D.K. Ray-Chaudhuri
1. Characterization of line-graphs. J. Combinatorial Theory
   3 (1967), 201-214.

F.S. Roberts
1. Representations of indifference relations. Ph.D. Thesis,
   Stanford University 1968.
2. On the boxicity and cubicity of a graph. (to appear)

A.C.M. van Rooij and H.S. Wilf
1. The interchange graph of a finite graph. Acta Math. Acad.
   Sci. Hungar. 16 (1965), 263-269.

M. Rosenfeld
1. On the total chromatic number of certain graphs. (to appear)

G. Sabidussi
1. Graph derivatives. Math. Z. 76 (1961), 385-401.

J. Sedláček
1. Some properties of interchange graphs. Theory of graphs and
   its applications. Smolenice 1963 (Prague 1964), 145-150.

S. Seshu and M. Reed
1. Linear graphs and electrical networks. Addison-Wesley, Reading 1961.

S.S. Shrikhande
1. On a characterization of the triangular association scheme.
   Ann. Math. Stat. 30 (1959), 39-47.
2. The uniqueness of the L_2 association scheme. Ann. Math.
   Stat. 30 (1959), 781-798.

E. Steinitz
1. Polyeder und Raumeinteilungen. Enzyk. Math. Wiss. 3AB12,
   pp. 1-139 (1922).

E. Steinitz and H. Rademacher
1. Vorlesungen über die Theoie der Polyeder. Springer, Berlin
   1934.

G. Wegner
1. Eigenschaften der Nerven homologisch-einfacher Familien.
   Ph.D. Thesis, University of Göttingen (1967).

H. Whitney
1. Congruent graphs and the connectivity of graphs. Amer. J.
   Math. 54 (1932), 150-168.

A.A. Zykov
  1. Problem 12. <u>Beiträge zur Graphentheorie</u>. (Edited by H. Sachs,
      H.-J. Voss and H. Walther.) Teubner, Leipzig 1968. page 228.

A MANY-FACETTED PROBLEM OF ZARANKIEWICZ

Richard K. Guy, University of Calgary

1. **Introduction.** Zarankiewicz [59] posed a problem, which naturally generalizes to that of finding the least $k = k_{i,j}(m,n)$ so that an $m \times n$ matrix, containing k ones and $mn - k$ zeros, no matter how distributed, contains an $i \times j$ submatrix consisting entirely of ones. We assume $2 \le i \le m$, $2 \le j \le n$, and write $k_i(m,n)$ if $i = j$, $k_{i,j}(n)$ if $m = n$, and omit one or both suffixes or arguments if the context is clear.

2. **Asymptotic Results.** Hartman et al. [29] obtained bounds for $k_2(n)$, which were improved by Kővári et al. [36], who showed that

$$\text{(1)} \qquad \lim_{n \to \infty} n^{-3/2} k_2(n) = 1,$$

and gave the exact result

$$\text{(2)} \qquad k_2(p^2 + p, p^2) = p^3 + p^2 + 1$$

in case p is prime. Hylteń-Cavallius [31] generalized their upper bound results to $k_{i,j}(m,n)$, and gave further asymptotic results for $i = 2$. Reiman [43] showed that there was equality in

$$\text{(3)} \qquad k_2(m,n) \le \tfrac{1}{2}\{m + (m^2 + 4mn(n-1))^{1/2}\} + 1$$

in infinitely many cases, e.g. in (2) and in

$$\text{(4)} \qquad k_2(p^2 + p + 1, p^2 + p + 1) = p^3 + 2p^2 + 2p + 2,$$

where p is a prime power, and noted the connection with affine and projective planes. Znám [60, 61] made successive improvements in the upper bound, and with the author [19] gave

Theorem 1. If $3 \le i \le m$, $3 \le j \le n$ and $n \ll m^i$, then $k_{i,j}(m,n) \le 1 + [nu]$, where $u = v + \tfrac{1}{2}(i-1)$, $v = z + (i^2-1)/24z + (i^2-1)(i^2-9)/1920z^3 + (i^2-1)(i^2-4)(i^2-25)/41472z^5$ and $nz^i = \cdot(j-1)m(m-1)\ldots(m-i+1)$.

In the case $i = 3$, we also showed that $k_{3,j}(m,n) \le 1 + [nu]$, where $u = z + \tfrac{1}{2}(i-1) + 1/3z$ and $nz^3 = (j-1)m(m-1)(m-2)$. For $i \ge 3$ it is difficult to obtain lower bounds, and the only result is that of Brown [3], who proved the first inequality in

$$\text{(5)} \qquad 2^{-1} \le n^{-5/3} k_3(n) \le 2^{-2/3} + \varepsilon,$$

thus partially confirming a conjecture of Kővári et al. [36], who gave the second inequality, and of Erdős [11]. However, the existence of the limit is still unproved, and nothing non-trivial appears to be known for $i \geq 4$.

3. <u>Exact Values</u>. For small values of the parameters, exact results are shown in Table 1 ($i = j = 2$), Table 2 ($i = 2$, $j = 3$), Table 3 ($i = 2$, $j = 4$), Table 4 ($i = j = 3$), Table 5 ($i = 3$, $j = 4$) and Table 6 ($i = j = 4$). The values in the left and upper borders of the tables are given by a theorem of Culik [8]:

<u>Theorem 2</u>. If $1 \leq i \leq m$ and $n \geq (j-1)\binom{m}{i}$, then

$$k_{i,j}(m,n) = (i-1)n + (j-1)\binom{m}{i} + 1.$$

The values between the two stepped lines in each table are given by a theorem of the author [17]:

<u>Theorem 3</u>. $k_{i,j}(m,n) = [\{(i^2-1)n + (j-1)\binom{m}{i}\}/i] + 1$, provided

$$(j-1)\binom{m}{i} + 1 \geq n \geq \ell(m,i,j).$$

The lower terminal for n has the approximate value $(j-1)\binom{m}{i}/(i+1)$. Its value has been made precise for small values of i.

Other values have been obtained by the methods of Section 12. The values of $k_3(n)$ for $n \leq 6$, $n = 7$, and $n = 8$ were found by Sierpiński [50], Brezeziński [51] and Culik [7].

4. <u>Graph-Theoretic Connections</u>. There are a number of ways in which matrices of zeros and ones may be connected with graphs. One is through the adjacency matrix, in which rows and columns represent vertices, and edges are represented by two elements 1 at the inter-sections of the rows and columns appropriate to the two adjacent ver-tices. All other elements are zero. With the usual meaning of graph, the matrix is symmetric with zeros on the leading diagonal. This is somewhat restricted in its application here, but examples can be giv-en. The adjacency matrix of K_m, the complete graph on m vertices, shows that $k_{i,m-i+1}(m) \geq m^2 - m + 1$; in fact, equality holds. It was noted by T.A. Jenkyns [17] that the adjacency matrix of the line graph of K_5 (i.e., the complement of the Petersen graph), shows that $k_3(10) \geq 61$; again equality holds.

A type of adjacency matrix which has more application here is that for bipartite graphs; instead of the rows and columns represent-ing the same vertices, they represent two distinct sets of vertices. Such matrices need not be symmetric, nor even square. We shall see, in Section 7, connections between such matrices and some cases of the Ramsey problem. Meanwhile, some other examples are provided by the edges of the 3-cube, which show that $k_{2,3}(4) \geq 13$; and by Figures 1 and 2 which show that $k_{2,3}(5,6) \geq 19$ and $k_3(6) \geq 27$.

Another matrix associated with a graph is its incidence matrix, where the rows and columns correspond to vertices and edges. Since an edge is incident with only 2 vertices, there are just 2 ones in each column. This is again restrictive in application here, but we will generalize the idea shortly. Even in the restricted form, the

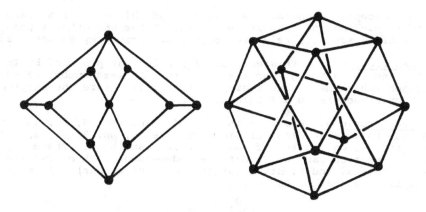

Figure 1   Figure 2

incidence matrix of K_m shows that $k_2(m, \frac{1}{2}m(m-1)) \geq m^2 - m + 1$, a
particular case of Theorem 2. More generally, we use the notion of
i-graph, in which the 'edges' are i-edges i.e. subsets of the vertices of cardinal i. Culik's theorem now follows on counting the
number of i-edges in K_m^i, the complete i-graph on m vertices,
where no edge may be counted more than j-1 times. The 'neighbor-
ing' Theorem 3 leads us to Section 5. Before leaving this one, note
that an incidence matrix is also appropriate to a projective geometry, where rows and columns represent points and lines, or to a family of sets, where they represent sets and elements. Hall and Ryser
[23] relate incidence matrices to difference sets and block designs.

5. <u>Packing and Covering</u>. The general problem can be restated in
terms of i-graphs. A column containing c ones corresponds to a
complete i-graph, K_c^i, on c vertices (rows), regarded as a subgraph of the complete graph K_m^i. Our object is to decompose K_m^i into complete subgraphs K_c^i in such a way as to maximize Σc, the total number of vertices in the subgraphs. Note that if an i-edge occurs as many as j times in different K_c^i, then a grid (i x j
submatrix of ones) is formed, so we may use each i-edge at most j - 1
times. In general, we have a mixed packing problem, but even in the
simpler case where c is constant over the columns, little is known
except for small i and c. Theorem 2 is the case c = i; and
Theorem 3 corresponds to c = i + 1. The dual problem is the covering problem, which coincides with the packing problem when the packing is exact. The covering problem was solved completely for i = 2,
c = 3 by Fort and Hedlund [15], and one can deduce the solution of
the corresponding packing problem.

     For c = 4, and i = 2 and 3, work has been done by Hanani
[25, 27] and Schönheim [47, 48]. Schönheim and the author hope
shortly to complete the two remaining of the six cases in the second

problem.

The concepts of coarseness [16, 18] and thickness of a graph are related to packing and covering problems. These and other topological concepts are discussed, relative to complete graphs by Beineke [1].

6. <u>Turán Problems</u>. A related type of problem was proposed by Turán [57, 58, and see also 9], who asked for the smallest number of edges in a graph (which may be an i-graph) which will ensure the existence of a given structure as a subgraph.

7. <u>Ramsey Problems</u>. Ramsey's theorem [41, and see 13, 14, 10] has given rise to a large class of problems [see 46, pp. 38-46, and references there], only a few of which are relevant here. There is, however, a connection, and we report the independent work of two students, V. Chvátal and S. Niven (written communications). In a notation similar to that of Rado [12], write

(6)
$$\binom{m}{n} \to \binom{i}{j}^r$$

if any r-coloring of the mn edges of the complete bipartite graph $K(m,n)$ ensures the existence of a monochromatic subgraph $K_{i,j}$, the i vertices being a subset of the m and the j vertices a subset of the n. The symbol

(7)
$$\binom{m}{n} \to \binom{i_1 \mid i_2}{j_1 \mid j_2}^r$$

means that any r-coloring of $K(m,n)$ contains either a monochromatic subgraph K_{i_1, j_1} or a monochromatic K_{i_2, j_2}. The negation of (6) and (7) is denoted by replacing " $\to$ " by " $\not\to$ ". We relate this to the present problem by the adjacency matrix of the bipartite graph. Whereas before the absence or presence of an edge was represented by zero or one, we now regard these as the names of two colors. Chvátal and Niven obtain the following result.

<u>Theorem 4</u>.
$$\binom{r(i-1)\binom{r(j-1)+1}{j}+1}{r(j-1)+1} \to \binom{i}{j}^r .$$

Niven also showed that this result is best possible in the sense that

(8)
$$\binom{r(i-1)\binom{r(j-1)+1}{j}}{r(j-1)+1} \not\to \binom{i}{j}^r .$$

For $r = 2$, there is a slightly stronger converse.

<u>Theorem 5</u>.
$$\binom{2(i-1)\binom{2j-1}{j}}{2j} \not\to \binom{i}{j}^2$$

Similarly he obtained

<u>Theorem 6</u>. $$\binom{2(i-1)\binom{2j-1}{j}}{2j} \not\to \binom{i|j}{j|i}^2, \quad \text{for } i \geq j,$$

and complete results in the following cases.

<u>Case 1</u>. $i = j = r = 2$.

$\binom{2}{x} \not\to \binom{2}{2}^2$, $x \geq 2$; $\binom{3}{7} \to \binom{2}{2}^2$; $\binom{4}{6} \not\to \binom{2}{2}^2$; $\binom{5}{5} \to \binom{2}{2}^2$.

From Table 1, $k_2(3,7) = 11$. By the pigeon-hole principle, the 21 entries in a 3×7 matrix must include at least 11 ones, or 11 zeros. Similarly $k_2(5,5) = 13 = (5^2 + 1)/2$. On the other hand, $k_2(4,6) = 13$ and $2(13 - 1) = 4 \times 6$, and it is possible to construct a matrix (Figure 3) which contains no 2×2 submatrix of either ones of zeros.

$$\begin{array}{cccccc} 1&1&1&0&0&0\\ 1&0&0&1&1&0\\ 0&1&0&1&0&1\\ 0&0&1&0&1&1 \end{array}$$

Figure 3

$$\begin{array}{cccccccccc} 0&0&1&0&0&1&0&1&1&1\\ 0&1&0&0&1&0&1&1&1&0\\ 1&0&0&1&0&1&1&1&0&0\\ 0&1&0&1&1&1&0&0&0&1\\ 1&0&1&1&1&0&0&0&1&0\\ 1&1&1&0&0&0&1&0&0&1 \end{array}$$

Figure 4

<u>Case 2</u>. $i = r = 2$, $j = 3$. The results

$\binom{2}{x} \not\to \binom{2}{3}^2$, $x \geq 3$; $\binom{3}{13} \to \binom{2}{3}^2$; $\binom{4}{12} \not\to \binom{2|3}{3|2}^2$;

$\binom{5}{11} \to \binom{2}{3}^2$; $\binom{6}{10} \not\to \binom{2|3}{3|2}^2$; $\binom{7}{9} \to \binom{2}{3}^2$ and $\binom{8}{8} \not\to \binom{2|3}{3|2}^2$,

may be compared with Table 2. For example, $k_{2,3}(7,9) = 32 = (63+1)/2$, while $k_{2,3}(6,10) = 31$, $2(31-1) = 6 \times 10$ and Figure 4 contains no 2×3 submatrix of zeros or of ones.

<u>Case 3</u>. $i = r = 2$, $j = 4$.

$\binom{2}{x} \not\to \binom{2}{4}^2$, $x \geq 4$; $\binom{3}{19} \to \binom{2}{4}^2$; $\binom{4}{18} \not\to \binom{2|4}{4|2}^2$;

$\binom{5}{15} \to \binom{2}{4}^2$; $\binom{8}{14} \not\to \binom{2|4}{4|2}^2$; $\binom{9}{13} \to \binom{2}{4}^2$; $\binom{12}{12} \not\to \binom{2|4}{4|2}^2$.

Compare with Table 3. Note that $\binom{5}{11} \to \binom{2}{3}^2$ implies $\binom{5}{11} \to \binom{2|3}{3|2}^2$,

but $\binom{5}{11} \not\to \binom{3}{2}^2$. In fact $k_{2,3}(5,11) = 27 = (55+1)/2$, but $k_{2,3}(11,5) = 33$.

Similar results may be obtained by studying Tables 4, 5 and 6. Results for $r = 3$ may be obtained by examining the tables for points where k just exceeds $mn/3$. For example Figure 5 shows both

$k_2(6,15) = 31$ (on reading the twos as zeros) and $\binom{6}{15} \nmid \binom{2}{2}^3$. Similarly Figure 6 shows $k_2(9,12) = 37$ and $\binom{9}{12} \nmid \binom{2}{2}^3$.

```
0 2 1 0 0 2 2 1 1 1 0 2 2 1 0        1 1 1 0 2 0 0 0 1 2 2 2
2 0 2 1 0 0 2 2 1 0 1 0 1 2 1        1 0 2 1 0 1 2 0 2 1 0 2
1 2 0 1 1 0 0 2 2 2 0 1 0 1 2        1 2 0 2 1 2 1 0 0 0 1 2
0 1 2 2 1 1 0 0 2 1 2 0 2 0 1        0 1 2 2 1 1 0 2 0 2 0 1
1 0 1 2 2 1 1 0 0 2 1 2 0 2 0        0 2 1 1 0 0 1 2 2 0 2 1
2 1 0 0 2 2 1 1 0 0 2 1 1 0 2        2 1 0 1 0 2 0 1 2 2 1 0
                                     2 0 1 2 1 0 2 1 0 1 2 0
                                     2 2 2 0 2 1 1 1 1 0 0 0
                                     0 0 0 0 2 2 2 2 1 1 1 1
```

       Figure 5   Figure 6

If we adjoin $j - 2$ replicas of its 15 columns to Figure 5, and similarly for Figure 6, Scott Niven obtains

$$\binom{6}{15j-15} \nmid \binom{2}{j}^3 \quad \text{and} \quad \binom{9}{12j-12} \nmid \binom{2}{j}^3 .$$

Also

$$\binom{12}{9j-9} \nmid \binom{2}{j}^3 \quad \text{and} \quad \binom{15}{6j-6} \nmid \binom{2}{j}^3 .$$

8. __Tournaments.__ Figure 7, which has been modified to illustrate the present problem, was recently produced by Kotzig (written communication) in quite a different context. From Kotzig's earlier work [34, 35] it is known that if p is prime and of the form $4q-1$, then there is a tournament between n players such that each game (oriented edge) belongs to exactly q cycles of length 3. The present example ($q = 4$, $n = 15$) is the first to show that n is not necessarily prime. The tournament represented by this matrix is depicted in Figure 8.

```
0 0 0 1 0 1 1 0 1 1 1 0 1 0 0
1 0 0 0 1 0 1 0 0 1 1 1 0 1 0
1 1 0 0 0 1 0 0 0 0 1 1 1 0 1
0 1 1 0 0 0 1 0 1 0 0 1 1 1 0
1 0 1 1 0 0 0 0 1 0 0 1 1 1 1
0 1 0 1 1 0 0 0 1 0 1 0 0 1 1
0 0 1 0 1 1 0 0 1 1 0 1 0 0 1
1 1 1 1 1 1 1 0 0 0 0 0 0 0 0
0 1 1 0 1 0 0 1 0 1 1 0 1 0 0
0 0 1 1 0 1 0 1 0 0 1 1 0 1 0
0 0 0 1 1 0 1 1 0 0 0 1 1 0 1
1 0 0 0 1 1 0 1 1 0 0 0 1 1 0
0 1 0 0 0 1 1 1 0 1 0 0 0 1 1
1 0 1 0 0 0 1 1 1 0 1 0 0 0 1
1 1 0 1 0 0 0 1 1 1 0 1 0 0 0
```

                    Figure 7

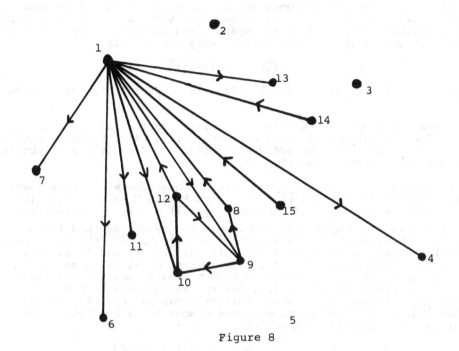

Figure 8

Only one-seventh of the games are indicated; the remainder are obtained by rotating the diagram through multiples of $2\pi/7$ about
player 8. Our present interest in Figure 7 is that it serves to show
that $k_{2,4}(15) \geq 106$; in fact equality holds.

9. <u>Steiner Triples</u>. The original problem [54] is classical, having
been solved by Reiss [44] and Moore [38]. Netto [39, pp. 202-227]
gives a good account of earlier work and Skolem [53] and Hanani[24]
have contributed to a complete solution. They pervade the whole of
combinatorics, and their connections with finite geometries [21],
with error-correcting codes [2, 49] and with block designs [25-28]
are well known. See Ryser [46, pp. 96-130] and Hall [22, pp. 100-119,
223-251] for references to the extnesive bibliography on this last
topic. More surprising is their occurrence in Ringel's beautiful
work [45] on the toroidal thickness of the complete graph. We are
interested in them for their own sake, and for their generalizations,
about which there is still to be discovered (compare Section 5).

The players among the first seven in Kotzig's tournament (Figure
7) who are beaten by these first seven (467, 157, 126, 237, 134, 245,
356) form a Steiner system, and in fact the first seven rows and sev-
en columns of the matrix demonstrate that $k_2(7) \geq 22$. It is again
easy to show equality. In the language of Section 5, they exhibit
a packing (or covering, since it is exact) of 7 triangles, K_3, in
K_7. If, in this 7 x 7 matrix, we interchange zeros and ones, we also have $k_{2,3}(7) \geq 29$; again there is equaltiy.

The row numbers of the ones (or zeros, or twos) in the columns of Figure 6 (123, 146, 157, 256, 347, 248, 358, 678, 189, 279, 369, 459) furnish another example of such a system, and more generally the existence of Steiner triple systems establishes the formula

$$(9) \qquad k_{2,j}(m, (j-1)\tbinom{m}{2}/3) = (j-1)\tbinom{m}{2} + 1,$$

provided $m \equiv 1$ or 3, modulo 6. This is a special case of Theorem 3.

Closely allied is Kirkman's schoolgirls problem [32, 33], recently solved completely by Ray-Chaudhuri and Wilson [42], to whose excellent bibliography should be added Netto [39, pp. 228-235], even though he misspells Kirkman's name throughout.

10. <u>Affine and Projective Planes</u>. Their connection with our present problem was noted by Reiman [43], who gave Figure 9, which shows (in effect) that $k_2(16,20) = 81$ and that $k_2(21) = 106$. Classical papers are by Bruck and Ryser [4] and Chowla and Ryser [5], and the connections with difference sets are shown in Hoffman [30] and Hall [20]. Hall's book [22, pp. 167-188] deals with this topic, and in successive chapters treats the related topics of orthogonal latin squares (see also Ryser [46, pp. 79-95]) and Hadamard matrices. Paley's paper [40] is the classic in this field, but unless one goes to the source, one is likely to overlook the two following papers, by Todd [56] and Coxeter [6], which show relations between what at first sight might appear to be quite different subjects.

```
0 1 0 0 0 0 1 0 1 0 0 0 0 0 0 1 1 0 0 0 │0
1 0 0 0 0 0 1 0 1 0 1 0 0 0 0 1 0 1 0 0 │0
0 0 0 1 1 0 0 0 0 1 0 0 1 0 0 1 0 1 0 0 │0
0 0 1 0 0 1 0 0 0 0 0 1 1 0 0 0 1 0 0 0 │0
0 0 1 0 1 0 0 0 0 1 0 0 0 0 0 1 0 1 0 0 │0
0 0 0 1 0 1 0 0 1 0 0 0 0 0 0 1 0 0 1 0 0 │0
1 0 0 0 0 0 1 0 0 0 0 1 0 1 0 0 0 1 0 0 │0
0 1 0 0 0 0 0 1 0 0 1 0 1 0 1 0 0 0 1 0 0 │0
1 0 0 0 0 1 0 0 0 0 1 0 0 0 0 1 0 0 1 0 │0
0 1 0 0 1 0 0 0 0 0 0 1 0 0 1 0 0 0 1 0 │0
0 0 1 0 0 0 0 1 1 0 0 0 0 1 0 0 0 0 1 0 │0
0 0 0 1 0 0 1 0 0 1 0 0 1 0 0 0 0 0 1 0 │0
0 0 0 1 0 0 0 1 0 0 0 1 0 0 0 1 0 0 0 1 │0
0 0 1 0 0 0 1 0 0 0 1 0 0 0 1 0 0 0 0 1 │0
0 1 0 0 0 1 0 0 0 1 0 0 0 1 0 0 0 0 0 1 │0
1 0 0 0 1 0 0 0 1 0 0 0 1 0 0 0 0 0 0 1 │0
1 1 1 1 0 0 0 0 0 0 0 0 0 0 0 0 0 0 0 0 │1
0 0 0 0 1 1 1 1 0 0 0 0 0 0 0 0 0 0 0 0 │1
0 0 0 0 0 0 0 0 1 1 1 1 0 0 0 0 0 0 0 0 │1
0 0 0 0 0 0 0 0 0 0 0 0 1 1 1 1 0 0 0 0 │1
0 0 0 0 0 0 0 0 0 0 0 0 0 0 0 0 1 1 1 1 │1
```

Figure 9

11. <u>Difference Sets</u>. These originate with Singer [52] and their importance was not immediately appreciated. However, there is now an extensive literature; see Hall [22, pp. 120-166] and Ryser [46, pp. 131-141] for references, and the paper of Emma Lehmer [37] and Storer's book [55] for the further connection with power residues and

cyclotomy. A simple example of a difference set is $\{0,1,5\}$, which has, for its differences of distinct members, representatives of each of the non-zero residue classes, modulo 7. It will be noted that additions to the members of the set generate the Steiner triples 126, 237, 341, ... mentioned in Section 9 as occurring in Kotzig's tournament; and which also form the lines of Fano's configuration (Figure 10), the projective plane of order 2.

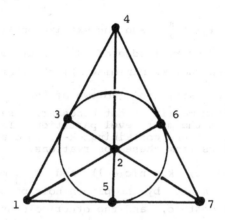

Figure 10

```
1 1 0 1 0 0 0 0 0 1 0 0 0
0 1 1 0 1 0 0 0 0 0 1 0 0
0 0 1 1 0 1 0 0 0 0 0 1 0
0 0 0 1 1 0 1 0 0 0 0 0 1
1 0 0 0 1 1 0 1 0 0 0 0 0
0 1 0 0 0 1 1 0 1 0 0 0 0
0 0 1 0 0 0 1 1 0 1 0 0 0
0 0 0 1 0 0 0 1 1 0 1 0 0
0 0 0 0 1 0 0 0 1 1 0 1 0
0 0 0 0 0 1 0 0 0 1 1 0 1
1 0 0 0 0 0 1 0 0 0 1 1 0
0 1 0 0 0 0 0 1 0 0 0 1 1
1 0 1 0 0 0 0 0 1 0 0 0 1
```

Figure 11

The matrix in Figure 11 is the incidence matrix of the projective plane of order 3, generated by the difference set $\{0,2,3,7\}$, and shows, effectively, that $k_2(13) = 57$. On omitting one column (line at infinity) and the 4 rows which 'intersect' it (ideal points), leaving the affine plane, we have $k_2(9,12) = 37$, which we have already seen from Figure 6. If we interchange zeros and ones in Figure 11, we have $k_{2,7}(13) = 118$.

12. <u>Methods and Examples</u>. This concluding section is intended to point to a need for further ideas, rather than to display a range of tools. The methods so far used are limited to counting arguments, generalizations of Dirichlet's pigeon-hole principle, and often tedious combinations of these with parity and other congruence arguments.

A. Suppose the column sum (number of ones) in column g is c_g, $1 \leq g \leq m$. Then the number of i-edges (see Section 5) in that column is $\binom{c_g}{i}$, and in all columns is $\Sigma_g \binom{c_g}{i}$. If this exceeds $(j-1)\binom{m}{i}$, then, by the pigeon-hole principle, there are j coincident (i.e., occupying the same i rows) i-edges, forming a grid (i.e. an $i \times j$ submatrix of ones).

B. If $\Sigma_g \binom{c_g}{i} = (j-1)\binom{m}{i}$ and the matrix contains no grid, it is said to be colmax, and every i-edge of K_m^i occurs exactly $j-1$ times. It can be seen that we minimize $\Sigma_g \binom{c_g}{i}$, for a total $\Sigma_g c_g$, by taking the 'most level partition' (no two of the c_g differing by more than 1). A colmax matrix is not necessarily saturated, but is so if its columns form the most level partition. In this context, a saturated matrix [43] is one containing $k-1$ ones and no grid. Culik's theorem follows from these observations.

C. If, for some c, $k < n(c + 1)$ and $k_{i,j}(m,n-1) \leq k - c$, then $k_{i,j}(m,n) \leq k$, since, by the pigeon-hole principle, at least one column sum is at most c, and the other columns contain a grid.

D. If an r-row (i.e. a row containing r ones) meets (has a one in common with) columns with sums c_g, $1 \leq g \leq r$, and

$$\sum_{g=1}^{r} \binom{c_g-1}{i-1} > (j-1)\binom{m-1}{i-1},$$

then the pigeon-hole principle ensures that at least j $(i-1)$-tuples coincide and form a grid with the r-row.

E. If one is packing K_m^i with specimens of K_c^i, $m > c > i$, then in general the number of i-edges available at a vertex of K_m^i (row of the matrix), i.e. $\binom{m-1}{i-1}$, is not an exact multiple of the number occurring in K_c^i, namely $\binom{c-1}{i-1}$. Hence it is possible to infer that a certain number of i-edges will be 'wasted' at each vertex, so a saturated matrix need not be colmax. This argument is used at a vertex (1-edge), but may also apply at a 2-edge if $\binom{c-2}{i-2} \nmid \binom{m-2}{i-2}$, and more generally at an e-edge, $1 \leq e \leq i$, if $\binom{c-e}{i-e} \nmid \binom{m-e}{i-e}$. Also, one may be able to combine information obtained sucessively from various values of e.

In these arguments we may interchange m and n, i and j, row and column. For brevity, we refer to these arguments by the capital letters, adding a prime if the argument is transposed. For example, Figures 3-7, 9 and 11 yield inequalities for special values of k. The reverse inequalities are given by A.

Example 1. $k_{4,2}(8,n) = 5n + 1$, $6 \leq n \leq 8$.

We wish to answer the question: how many K_5^4 can be packed in K_8^4? It is easier to answer the complementary one: how many K_3^2 in K_8^2? Note that $8 - 5 = 3$ and $(2 \times 5) - 8 = 2$. The graph K_8 contains 28 edges, 7 at a vertex, so by E, at most 3 triangles can be packed at a vertex; 8 triangles in all. Such a packing is possible and is represented by the zeros in Figure 12, which show that $k_2(8) \geq 25$. In this case an argument stronger than A (E will serve) is needed to show equality. The ones in Figure 12 show that $k_{4,2}(8,n) \geq 5n + 1$, $2 \leq n \leq 8$. To see that equality holds for $n = 6$, use C' with $c = 3$ and $k_{4,2}(7,6) = 28 = 31-3$. The results for $n = 7$ and 8 now follow inductively by C' with $c = 4$.

```
1 1 1 1 1 0 0 0
1 1 0 0 0 1 1 1
1 0 1 1 0 1 1 0          1 0 0 1 1 1 0 1 0 1
1 0 1 0 1 0 1 1          1 1 0 0 1 1 1 0 1 0
1 0 0 1 1 1 0 1          1 1 1 0 0 0 1 1 0 1
0 1 1 0 1 1 0 1          0 1 1 1 0 1 0 1 1 0
0 1 1 1 0 0 1 1          0 0 1 1 1 0 1 0 1 1
0 1 0 1 1 1 1 0
```

| Figure 12 | Figure 13 |

Example 2. $k_{2,4}(5,9) = 28$.

Figure 13 shows that $k_{2,4}(5,n) > 3n$, $4 \leq n \leq 10$. Suppose that $k(5,9) > 28$ so that a saturated matrix contains at least 28 ones. By B, the columns would be $3^8 4$ (in a partition, the indices denote repetition, not exponentiation), and since $8\binom{3}{2} + \binom{4}{2} = (4-1)\binom{5}{2}$, the matrix is colmax. However, as in E, it is not possible to pack 8 triangles and a (complete) quadrangle in K_5, using the edges not more than three times, since 12 edges are available at a vertex, of which the quadrangle takes an odd number, 3, so some edges are 'wasted'.

Example 3. $k_{2,4}(6,9) = 33$.

Figure 14 shows $k > 32$. If $k = 34$, there would be a matrix with 33 ones and no grid. Such a matrix would contain a 5-row (or less). Delete this, and note that C' with $c = 5$ applies, by Example 2, since $28 = 33 - 5$.

```
                              0 1 1 1 1 1 1 1
0 0 1 0 1 1 1 1 1             1 0 1 1 1 0 1 0
1 0 0 1 0 1 1 1 0             1 1 0 1 0 1 1 0
1 1 0 0 1 0 1 1 0             1 1 1 0 1 1 0 0
1 1 1 0 0 1 0 1 0             1 1 0 1 1 0 1 1
1 1 1 1 0 0 1 0 1             1 0 1 1 0 1 0 1
0 1 1 1 1 0 0 1 0             1 1 1 0 0 0 1 1
```

| Figure 14 | Figure 15 |

<u>Example 4</u>. $k_3(7,8) = 38$.

Figure 15 shows $k > 37$. By A, we could have a grid-free matrix containing 38 ones only if the columns were $5^6 4^2$. Now it is not possible to have 6 5-columns, nor even 5, since, if 2 5-columns coincide (occupy the same 5 rows), no other column can contain more than 4 ones. If 2 5-columns have just 4 rows in common, any other 5-column has at most 2 ones in these rows, i.e. 3 in the remaining 3 rows. Three such columns form a grid. Consider the 7 x 5 matrix formed by 5 5-columns. By the pigeon-hole principle, 4 of its rows contain at least 16 ones between them. Consider the 4 x 5 matrix formed from these rows. One column contains 4 ones, and at most one column, else 2 5-columns in the original matrix would have 4 rows in common. So the other 4 columns contain 3 ones and a zero, the zeros being in different rows, as in the first 4 rows of Figure 16. It is not possible to complete the last 3 rows with no two 5-columns having 4 rows in common, a grid must be formed.

```
1 1 1 1 0          1 1 0 1 1 1 1 1
1 1 1 0 1          1 1 1 1 1 1 0 0
1 1 0 1 1          1 1 1 1 0 0 1 1
1 0 1 1 1          1 0 1 0 1 1 1 1
1 0                1 1 1 0 1 0 1 0
0 1                0 1 1 1 0 1 0 1
0 1
```

Figure 16   Figure 17

<u>Example 5</u>. $k_{3,4}(6,8) = 36$.

Figure 17 shows $k > 35$. If a grid-free matrix with 36 ones contained a 6-column, it would form a 3,4-grid with the other seven columns, since $k_3(6,7) = 30$. There is no 3-column by C with c = 3, so the columns are $5^4 4^4$. A 7-row would meet columns $5^3 4^3$ or more, and form a grid with the other 5 rows, since $k_{2,4}(5,7) = 24$, so the rows are 6^6. If 2 6-rows coincide, any other 6-row forms a grid with them. If 2 6-rows overlap in 5 columns, each row has at most 3 ones in these columns, and hence ones in all the other three columns. These columns then form a grid with any other 5-column. So no pair of 6-rows overlap in more than 4 columns. At most 4 6-rows can satisfy such a condition.

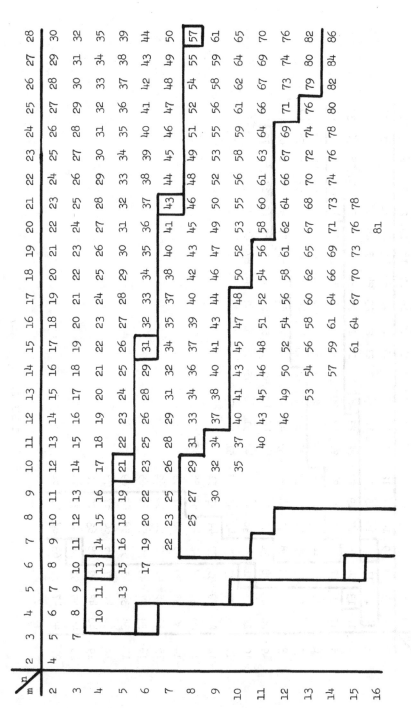

TABLE 1. $k_2(m,n)$

TABLE 2. $k_{2,3}(m,n)$

m \ n	3	4	5	6	7	8	9	10	11	12	13	14	15	16	17	18	19	20	21	22	23	24
2	6	7	8	9	10	11	12	13	14	15	16	17	18	19	20	21	22	23	24	25	26	27
3	8	10	11	13	14	15	16	17	18	19	20	21	22	23	24	25	26	27	28	29	30	31
4	10	13	14	16	17	19	20	22	23	25	26	27	28	29	30	31	32	33	34	35	36	37
5	12	15	17	19	21	23	24	26	27	29	30	32	33	35	36	38	39	41	42	43	44	45
6	14	17	20	22	25	26	28	31	32	34	35	37	38	40	41	43	44	46	47	49	50	52
7	16	19	23	25	29	30	32	35	37	39	40	43	44	46	47	49	50	52	53	55	56	58
8	18	21	25	28	32	34	36	39	41	43	45	47	49	51	53	55	57	59	60	62	63	65
9	20	23	28	31	35	37	40	43	46	47	50	52	55	57	58	60	62	65	67	69	70	73
10	22	25	31	34	38	41	43	47	51	52	55	57	61	62	64	67	68	71	73			79
11	24	27	33	37	41	45	47	51	56	57												
12	26	29	35	39	44	49	51	54														
13	28	31	37	42	47	53	55	57														
14	30	33	39	45	50	57	59															
15	32	35	41	47	52	59	62															
16	34	37	43	50	55	62	65															
17	36	39	45	53	58	65	69															
18	38	41	47	55	60	67	73															
19	40	43	49	58	63	70	76															
20	42	45	51	61	66	73	79															
21	44	47	53	63	68	75	82															

m \ n	4	5	6	7	8	9	10	11	12	13	14	15	16	17	18	19	20	21	22	23	24
2	8	9	10	11	12	13	14	15	16	17	18	19	20	21	22	23	24	25	26	27	28
3	11	13	14	16	17	19	20	21	22	23	24	25	26	27	28	29	30	31	32	33	34
4	14	17	18	20	22	23	25	26	28	29	31	32	34	35	37	38	39	40	41	42	43
5	17	21	22	24	27	28	31	32	34	35	37	38	40	41	43	44	46	47	49	50	52
6	20	24	26	28	31	33	36	38	40	41	43	45	47	49	50	52	53	55	56	58	59
7	23	27	30	32	36	38	41	43	45	47	50	52	54	56	58	59	61	64	65	67	68
8	26	30	34	36	41	43	46	49	51	53	57	58	60	63	65	67	69	71	73	75	77
9	29	33	38	40	45	48	51	55	57	59											
10	32	36	42	44	49	53	56	61													
11	35	39	46	48	53	58	61	67													
12	38	42	49	52	57																
13	41	45	53	56	61																
14	44	48	57	60	65																
15	47	51	61	64	69						106										
16	50	54	64	68	73																

TABLE 3. $k_{2,4}(m,n)$

m \ n	3	4	5	6	7	8	9	10	11	12	13	14	15	16	17	18	19	20	21	22	23
3	9	11	13	15	17	19	21	23	25	27	29	31	33	35	37	39	41	43	45	47	49
4		14	17	19	22	25	27	29	31	33	35	37	39	41	43	45	47	49	51	53	55
5			21	23	26	29	31	34	37	39	42	45	47	50	53	55	58	61	63	65	67
6				27	30	33	37	40	43	46	49	51	54	57	59	62	65	67	70	73	75
7					34	38	41	45	48	51	54	57	61	64	67	70	73	76	79	82	85
8						43	46	51	54	58	61	65									
9							50	55	60	65											
10								61													

TABLE 4. $k_{3}(m,n)$

m\n	4	5	6	7	8	9	10	11	12	13	14	15	16	17	18	19	20	21
3	12	14	16	18	20	22	24	26	28	30	32	34	36	38	40	42	44	46
4	15	18	21	23	26	29	31	34	37	39	41	43	45	47	49	51	53	55
5	18	22	26	28	31	34	37	40	43	45	48	51	53	56	59	61	64	67
6	21	26	31	33	36	40	43	47	51	53	57	61	63	66	69	71	74	77
7	24	29	34	38	41	45	49	53	57									
8	27	33	38	43	46													
9	30	37	42	47	51													
10	33	41	46	52	56													
11	36	44	49	56	61													
12	39	47	53	61	66													
13	42	50	57	65	71													
14	45	53	61	69	75													
15	48	56	64	73	80													
16	51	59	68	77	85													
17	54	62	72	81	90													
18	57	65	76	85	95													
19	60	68	79	89	100													

TABLE 5. $k_{3,4}(m,n)$

m\n	4	5	6	7	8	9	10	11	12	13	14	15	16	17	18	19	20	21
4	16	19	22	25	28	31	34	37	40	43	46	49	52	55	58	61	64	67
5		23	27	31	34	38	42	46	49	53	57	61	64	67	70	73	76	79
6			32	37	40	44	48	52	56	60	64	68	72	76	79	83	87	91
7				43	46	50	55	59	64	69	73	78	83	88	91	96	101	106
8					52	56	61											

TABLE 6. $k_4(m,n)$

REFERENCES

1. L.W. Beineke, Topological aspects of complete graphs, in G. Katona (ed.) Theory of Graphs, Akadémiai Kiadó, Budapest, 1968, 19-26.

2. E.R. Berlekamp, Algebraic Coding Theory, McGraw-Hill, 1968.

3. W.G. Brown, On graphs which do not contain a Thomsen graph, Canad. Math. Bull., 9 (1966), 281-285, MR 34 (1967), #81.

4. R.H. Bruck and H.J. Ryser, The non-existence of certain finite projective planes, Canad. J. Math., 1 (1949), 88-93. MR 10(1949), 319.

5. S. Chowla and H.J. Ryser, Combinatorial problems, Canad. J. Math., 2 (1950), 93-99. MR 11 (1950), 306.

6. H.S.M. Coxeter, Regular compound polytopes in more than four dimensions, J. Math. and Phys., 12 (1932-3), 334-345.

7. K. Culik, Poznámka k problému K. Zarankiewicze, Pracé Brnenské Základny CSAV, 26 (1955), 341-348.

8. K. Culik, Teilweise Lösung eines verallgemeinerten Problem von K. Zarankiewicz, Ann. Polon. Math., 3 (1956), 165-168. MR 18 (1957), 459.

9. P. Erdös, Extremal problems in graph theory, in M. Fiedler (ed.), Theory of Graphs and its Applications (Proc. Symp. Smolenice, (1963), Prague, 1964, 29-36. MR 31 (1966), #4735.

10. P. Erdös, Some applications of probability to graph theory and combinatorial problems, in M. Fiedler (ed.), Theory of Graphs and its Applications (Proc. Symp. Smolenice, 1963), Prague, 1964, 29-36, MR 30 (1965), #3459.

11. P. Erdös, On extremal problems of graphs and generalized graphs, Israel J. Math., 2 (1964), 183-190. MR 32 (1966), #1134.

12. P. Erdös and R. Rado, A partition calculus in set theory, Bull. Amer. Math. Soc., 62 (1956), 427-489. MR 18 (1957), 458.

13. P. Erdös and R. Rado, A combinatorial theorem, J. London Math. Soc., 25 (1950), 249-255. MR 13 (1952), 322.

14. P. Erdös and G. Szekeres, A combinatorial problem in geometry, Compositio Math., 2 (1935), 463-470.

15. M.K. Fort, Jr. and G.A. Hedlund, Minimal coverings of pairs by triples, Pacific J. Math., 8 (1958), 709-719. MR 21 (1960), #2595.

16. R.K. Guy, A coarseness conjecture of Erdös, J. Combinatorial Theory, 3 (1967), 38-42. MR 35 (1968), #78.

17. R.K. Guy, A problem of Zarankiewicz, in G. Katona (ed.), Theory of Graphs, Akadémiai Kiadó, Budapest, 1968, 119-150.

18. R.K. Guy and L.W. Beineke, The coarseness of the complete graph, Canad. J. Math., 20 (1968), 888-894. MR 37 (1969), #2633.

19. R.K. Guy and S. Znám, A problem of Zarankiewicz, in W.T. Tutte (ed.), Recent Progress in Combinatorics, Academic Press, 1969.

20. M. Hall, Cyclic projective planes, Duke Math. J., 14 (1947), 1079-1090. MR 9 (1948), 370.

21. M. Hall, Automorphisms of Steiner triple systems, I.B.M. J. Res. Develop., 4 (1960), 460-472, MR 23 (1962), #A1282.

22. M. Hall, Combinatorial Theory, Blaisdell, 1967. MR 37 (1969), #80.

23. M. Hall and H.J. Ryser, Cyclic incidence matrices, Canad. J. Math., 3 (1951), 495-502. MR 13 (1952), 312.

24. H. Hanani, A note on Steiner triple systems, Math. Scand., 8 (1960), 154-156. MR 23 (1962), #A2330.

25. H. Hanani, On quadruple systems, Canad. J. Math., 12 (1960), 145-157, MR 22 (1961), #2558.

26. H. Hanani, The existence and construction of balanced incomplete designs, Ann. Math. Statist., 32 (1961), 361-386. MR 29 (1965), #4161.

27. H. Hanani, On some tactical configurations, Canad. J. Math., 15 (1963), 702-722. MR 28 (1964), #1136.

28. H. Hanani and J. Schönheim, On Steiner systems, Israel J. Math., 2 (1964), 139-142. MR 31 (1966), #73. Addendum, ibid., 4 (1966), 144. MR 34 (1967), #72.

29. S. Hartman, J. Mycielski, and C. Ryll-Nardzewski, Colloq. Math., 3 (1954), 84-85.

30. A.J. Hoffman, Cyclic affine planes, Canad. J. Math., 4 (1952), 295-301. MR 14 (1953), 196.

31. C. Hylteń-Cavallius, On a combinatorical problem, Colloq. Math., 6 (1958), 59-65. MR 21 (1960), #1941.

32. T.P. Kirkman, On triads made with fifteen things, London Edinburgh and Dublin Phil. Mag., 37 (1850), 169-171.

33. T.P. Kirkman, On the puzzle of the fifteen young ladies, London Edinburgh and Dublin Phil. Mag., (4), 23 (1862), 198-204.

34. A. Kotzig, Cycles in a complete graph oriented in equilibrium, Mat. Fyz. Casopis, 16 (1966), 175-182. MR 34 (1967), #2491.

35. A. Kotzig, Des cycles dans les tournois, in P. Rosenstiehl (ed.), Théorie des Graphes, (Symp. Rome, 1966), Dunod, Paris, 1967,

203-208.

36. T. Kővári, V. Sós, and P. Turán, On a problem of K. Zarankiewicz, Colloq. Math., 3 (1954), 50-57. MR 16 (1955), 456.

37. E. Lehmer, On residue difference sets, Canad. J. Math., 5 (1953), 425-432. MR 15 (1954), 10.

38. E.H. Moore, Concerning triple systems, Math. Ann., 43 (1893), 271-285.

39. E. Netto, Lehrbuch der Combinatorik, Leipzig, 1927, Chelsea, 1958. MR 20 (1959), #1632.

40. R.E.A.C. Paley, On orthogonal matrices, J. Math. and Phys., 12 (1933), 311-320.

41. F.P. Ramsey, On a problem of formal logic, Proc. London Math. Soc. (2), 30 (1930), 264-286.

42. D.K. Ray-Chaudhuri and R.M. Wilson, Solution of Kirkman's school-girl problem, Proc. Amer. Math. Soc. Symp. on Combinatorics, Los Angeles, 1968.

43. I. Reiman, Uber ein Problem von K. Zarankiewicz, Acta Math. Acad. Sci. Hungar., 9 (1958), 269-279. MR 21 (1960), #63.

44. M. Reiss, Ueber eine Steinersche combinatorische Aufgabe welche in 45-sten Bande dieses Journals, Seite 181, gestellt worden ist, J. Reine Angew. Math., 56 (1859), 326-344.

45. G. Ringel, Die toroidale Dicke des vollständigen Graphen, Math. Z., 87 (1965), 19-26. MR 30 (1965), #2489.

46. H.J. Ryser, Combinatorial Mathematics, Carus Monograph #14, Math. Assoc. Amer. 1963. MR 27 (1964), #51.

47. J. Schönheim, On coverings, Pacific J. Math., 14 (1964), 1405-1411, MR 30 (1965), #1954.

48. J. Schönheim, On maximal systems of k-tuples, Studia Sci. Math. Hungar., 1 (1966), 363-368. MR 34 (1967), #2485.

49. J. Schönheim, On linear and non-linear simple error-correcting q-nary perfect codes, Information and Control, 12 (1968), 23-26.

50. W. Sierpiński, Sur un probléme concernant un réseau a 36 points, Ann. Polon. Math., 24 (1951), 173-174. MR 15 (1954), 594.

51. W. Sierpiński, Problems in the Theory of Numbers (trans. A. Sharma), Pergamon, 1964, 16. MR 30 (1965), #1078.

52. J. Singer, A theorem in finite projective geometry and some applications to number theory, Trans. Amer. Math. Soc., 43 (1938), 377-385.

53. T. Skolem, Some remarks on the triple systems of Steiner, Math. Scand., 6 (1958), 273-280. MR 21 (1960), #5582.

54. J. Steiner, Combinatorische Aufgabe, <u>J. Reine Angew. Math.</u>, 45 (1853), 181-182.

55. T. Storer, <u>Cyclotomy and Difference Sets</u>, Markham, 1967. MR 36 (1968), #128.

56. J.A. Todd, A combinatorial problem, <u>J. Math. and Phys.</u>, 12 (1933), 321-333.

57. P. Turán, Eine Extremalaufgabe aus der Graphentheorie, <u>Mat. Fiz. Lapok</u>, 48 (1941), 436-452. MR 8 (1947), 284.

58. P. Turán, On the theory of graphs, <u>Colloq. Math.</u>, 3 (1954), 19-30.

59. K. Zarankiewicz, Problem P101, <u>Colloq. Math.</u>, 2 (1951), 301.

60. S. Znám, On a combinatorical problem of K. Zarankiewicz, <u>Colloq. Math.</u>, 11 (1963), 81-84. MR 29 (1965), #37.

61. S. Znám, Two improvements of a result concerning a problem of K. Zarankiewicz, <u>Colloq. Math.</u>, 13 (1965), 255-258. MR 32 (1966), #7434.

GRAPH THEORY AND LIE ALGEBRA

Ronald C. Hamelink, Michigan State University

This paper presents two examples occurring in the classification of simple Lie algebras in which graphs occur. The first situation is well known in Lie algebra and the conditions imposed on the graphs make the determination of eligible graphs very easy; the second case is rather new and much is still to be learned about it.

Case I. In the classification of simple, finite dimensional Lie algebras over any algebraically closed field of characteristic zero, conditions are laid upon a class of graphs, called Dynkin diagrams. The determination of this class is essential to solving the algebra problem. The conditions on the graphs, G, in this class are the following:

    i.   G is a connected, finite graph.

    ii. G may have multiple edges, but not loops.

  iii. The number of edges meeting in a vertex is at most three.

  iv. G contains no circuits on 3 or more vertices.

       (⬤◯⬤ is not excluded.)

    v. If $v_1, v_2, \ldots, v_n$ is a set of vertices in G, v_i adjacent to v_{i+1} with exactly one edge between them, then the graph G' obtained by deleting $v_1, \ldots, v_n$ and inserting a new vertex v, with v adjacent to each vertex u in $G - \{v_1, \ldots, v_n\}$ with the same number of lines which previously connected u to the set $\{v_1, \ldots, v_n\}$ is again in this class.

  vi. No graph G contains a proper subgraph of the form

$$\bullet\!-\!\bullet\!\!=\!\!\bullet\!-\!\bullet$$

vii. In a graph of the form

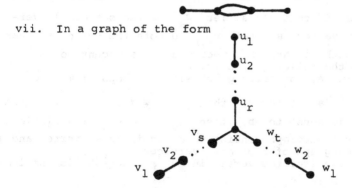

with $1 \le r \le s \le t$, either $r = s = 1$ and $t = 1,2,3,\ldots$ or $r = 1$, $s = 2$ and $t = 2,3$, or 4.

From this information the class of graphs can be completely determined. There are 3 infinite families.

1. The simple paths.

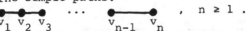

v_1 v_2 v_3 $\quad\quad v_{n-1}$ v_n , $n \geq 1$.

2. The paths with a double edge at one end

v_1 v_2 v_3 $\quad\quad v_{n-1}$ v_n , $n \geq 2$.

3. The paths beginning with ⟩–

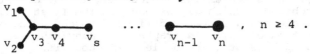

v_1

v_2 $\quad v_3$ v_4 $\quad v_s$ $\quad\quad v_{n-1}$ v_n , $n \geq 4$.

There are also 5 exceptional graphs in this class. They are

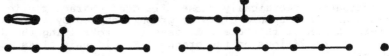

The classification theory for Lie algebras then shows that each of these graphs determines a unique simple Lie algebra with the exception of family 2, for which each graph with order ≥ 3 determines two non-isomorphic simple Lie algebras. The reader interested in the algebraic aspects of this classification should see [2].

Case II. This case is also concerned with the classification of simple Lie algebras. However, here the field is an algebraically closed field of characteristic 2. Several restrictive hypotheses are put on the algebra, mainly to compensate for the absence of characteristic zero, but also to make the class of graphs occurring one that can be determined. The reader interested in the algebraic details should see [1]. Here we will list the conditions imposed on the graphs by the Lie algebra and then proceed to determine the class of all such graphs.

For a graph G in this class, let V_a be the set of all vertices adjacent to a vertex a. Let $V_{a,b}$ denote the set of all vertices adjacent to both a and b, where a is adjacent to b. Then G satisfies the following conditions.

   i. G is a finite, connected graph without loops or multiple edges.

  ii. If a is adjacent to b, then $V_{a,b} \neq \emptyset$.

 iii. If a is adjacent to b, then $V_{a,b}$ is either a singleton, or induces a subgraph consisting of an isolated vertex and a complete subgraph of 2 or more vertices.

 iv. If c is the isolated vertex in $V_{a,b}$ then b is the isolated vertex in $V_{a,c}$.

  v. If b and c are both not adjacent to a then the isolated vertex in $V_{b,c}$ is also not adjacent to a.

 vi. The cardinality of $V_{a,b}$ is a constant for any two adjacent vertices, and the degree of a is $2|V_{a,b}|$.

From vi., it follows immediately that G is a strongly regular graph of even degree. Let $2n$ denote this degree and G_n denote any graph in this class of degree $2n$. We will define a pair (unordered) in V_a to be a set $\{x,y\} \subseteq V_a$ such that y is isolated in the section graph on $V_{a,x}$.

Lemma 1. The pairs partition V_a.

Proof. For any $x \in V_a$, $V_{a,x}$ contains a unique isolated vertex so each vertex is contained in one pair. Because the isolated vertex is unique and because y isolated in $V_{a,x}$ implies x isolated in $V_{a,y}$, we see that each vertex is contained in exactly one pair.

Lemma 2. Let $b \in V_a$, then each pair in V_a consists of one element from $V_{a,b}$ and one element from $V_a - V_{a,b}$.

Proof. Let $\{x,y\}$ be a pair in V_a. Assume both $x,y \in V_{a,b}$. Then y is adjacent to b and both y and b are adjacent to x. This contradicts y being isolated in $V_{a,x}$. So each vertex in $V_{a,b}$ is paired with a vertex in $V_a - V_{a,b}$, the fact that $|V_a| = 2|V_{a,b}|$ completes the argument.

We now pick an arbitrary, but fixed vertex a. Let $V_a = \{b_1, b_2, \ldots, b_n, c_1, c_2, \ldots, c_n\}$ where the notation is such that c_i is isolated in V_{a,b_i} and $V_{a,b_i} = \{b_1, b_2, \ldots, b_{i-1}, b_{i+1}, \ldots, b_n, c_i\}$. The structure of G_1 is readily seen to be

and all further consideration will be for G_n, $n \geq 3$. Let v be the vertex set of G_n, and $W = V - (V_a \cup \{a\})$.

Lemma 3. For $n \geq 3$, $w \neq \phi$.

Proof. Take pairs $\{b_1, c_1\}$ and $\{b_2, c_2\}$ in V_a and let d be the isolated vertex in V_{b_1, b_2}. Since a is also in V_{b_1, b_2} either $d \in W$ or $d = a$. Because $n \geq 3$, b_3 is also in V_{b_1, b_2}, and since a is adjacent to b_3, we see that $d = a$ is impossible.

For notational purposes, let $d_{i,j}$ denote the isolated vertex in V_{b_i, b_j}.

Lemma 4. $d_{i,j}$ is adjacent to c_i and c_j.

Proof. $V_{b_i} = \{a, c_i, b_1, \ldots, b_{i-1}, b_{i+1}, \ldots, b_n, d_{1,i}, d_{2,i}, \ldots, d_{i-1,i}, d_{i+1,i}, \ldots, d_{n,i}\}$. Because c_i is isolated in V_{a,b_i} it follows that $b_1,\ldots,b_{i-1},b_{i+1},\ldots,b_n$ are not in V_{b_i,c_i} so $V_{b_i,c_i} =$ $\{a_1,d_{1,i},\ldots,d_{i-1,i},d_{i+1,i},\ldots,d_{n,i}\}$. In particular $d_{ij} \in V_{c_i}$ and similarly $d_{ij} \in V_{c_j}$.

Lemma 5. d_{ij} is the isolated vertex in V_{c_i,c_j}.

Proof. This follows easily since we know $V_{c_i,c_j} = \{a,d_{ij}\} \cup \{c_k \mid k \neq i,j\}$ and a is adjacent to each c_k.

Lemma 6. d_{ij} is adjacent to $d_{k\ell}$ if and only if they have a common index.

Proof. Because $V_{b_i} = \{a,c_i\} \cup \{b_k \mid k \neq i\} \cup \{d_{ik} \mid k \neq i\}$ and $V_{b_i,b_j} = \{a,d_{ij}\} \cup \{b_k \mid k \neq i \text{ or } j\}$ with d_{ij} isolated it must follow that $V_{b_i,d_{ij}} = \{b_j c_i\} \cup \{d_{ik} \mid j \neq k \neq i\}$. Similarly $V_{b_j,d_{ij}} = \{b_i,c_j\} \cup \{d_{jk} \mid i \neq k \neq j\}$. Therefore it must follow that

$$V_{d_{ij}} = \{b_i,b_j,c_i,c_j\} \cup \{d_{rs} \mid \{r,s\} = \{i,k\} \text{ or } \{r,s\} = \{j,k\}, i \neq k \neq j\},$$

which shows the desired result.

The graphs G_n can now be determined with the aid of total graphs.

Theorem 1. G_n is isomorphic to $T(K_{n+1})$, the total graph of the complete graph on $n + 1$ points.

Proof. Denote the vertices of K_{n+1} by the integers $0,1,\ldots,n$ and the edges of K_{n+1} by the collection of 2 element subsets of these integers. Then $T(K_{n+1})$ has the vertex set

$$\{0,1,\ldots,n\} \cup \{\{i,j\} \mid 0 \le i \neq j \le n\}.$$

The edge set of $T(K_{n+1})$ is

$$\{\{i,j\} \mid 0 \le i \neq j \le n\} \cup \{\{i,\{i,j\}\} \mid 0 \le i \neq j \le n\} \cup \{\{\{i,j\}, \{j,k\}\} \mid i,j \text{ and } k \text{ distinct integers between } 0 \text{ and } n\}.$$

G_n has the vertex set $\{a\} \cup \{b_i \mid i = 1,\ldots,n\} \cup \{c_i \mid i = 1,\ldots,n\} \cup \{d_{ij} \mid 1 \le i \neq j \le n\}$. This is the complete set of vertices of G_n because each of these vertices is adjacent to only other vertices in this set and G_n is connected. Define the mapping α by

$$\alpha(a) = 0$$

$$\alpha(b_i) = i \qquad i = 1, \ldots, n \; ;$$

$$\alpha(c_i) = \{0, i\} \qquad i = 1, \ldots, n \; ;$$

$$\alpha(d_{ij}) = \{i, j\} \qquad 1 \leq i \neq j \leq n \; .$$

We can see that α is a one-to-one correspondence on the respective vertex sets. Further, Lemmas 2, 4, and 6 (using the notation a, b_i, c_i, and d_{ij} for vertices of G_n) show that α induces a 1-1 map from the edge set of G_n into the edge set of $T(K_{n+1})$. Now the fact that vertices in G_n and in $T(K_{n+1})$ have the same degree completes the theorem.

Results similar to those mentioned in case I now follow. Each of G_n, $n \geq 3$, determines 2 non-isomorphic simple Lie algebras. G_1 determines 3 non-isomorphic simple Lie algebras. For algebraic reasons the graph G_2 determines no simple Lie algebras, and so is excluded from the class of eligible graphs.

In concluding the discussion of this case it must be said that the algebraic condition needed to yield this class of graphs is rather restrictive. It would be of interest to find the class of graphs determined when we replace hypothesis iii. by

iii'. If a is adjacent to b then the subgraph induced by $V_{a,b}$ consists of an isolated vertex, and a connected component.

This change will enlarge the class of graphs since one example of this type is known.

REFERENCES

1. R. Hamelink, Lie algebras of characteristic 2, _Trans. Amer. Math. Soc._, to appear.

2. N. Jacobson, _Lie algebra_, Interscience Publishers, New York, 1962.

MATROIDS VERSUS GRAPHS[1]

Frank Harary, The University of Michigan
Dominic Welsh, Merton College, Oxford University

Almost 200 years elapsed between the first recorded discoveries of graph theory and matroid theory, for Euler characterized "eulerian graphs" in 1737 and Whitney developed axiom systems for matroids in 1935. Whitney's paper lay dormant until Tutte and Rado wrote independently on the subject in 1957. Since then interest in matroid theory has been accelerating rapidly, along with all other aspects of combinatorial theory. It can be argued that matroid theory serves to unify several different areas of combinatorics. Nevertheless, research in this area has been deliberately avoided by most graph theorists. This can be explained partly because the pictorial intuitive appeal of graph theory is not present in matroids and partly because of the proliferation of axiom systems for matroids, which can be confusing.

Our objective is to begin to remove this prejudice from the minds and hearts of graph theorists. One of us has recently written a book on graph theory [5] and the other is preparing a monograph on matroid theory [21]. The notations and terminology in this article will be consistent with that used in both of these books.

The various aspects of matroid theory, including undefined terms, axioms, concepts, and theorems, will always be linked to the corresponding aspects of graph theory so that it should become apparent to the students of graph theory, to whom this article is addressed, that matroid theory constitutes a natural generalization of graph theory. Such an analysis could also be carried out as a generalization of the concept of independence in vector spaces, but that will not be included here.

It is not only convenient but essential that the reader have a knowledge of the fundamentals of graph theory as given, for example, in the book [5]. It is neither desirable nor necessary for reading and understanding this expository development to assume that the reader is an expert on the theory of matroids. By an abuse of language, we use the word graph in this paper to mean a 'pseudograph', in which both loops and multiple lines are permitted.

In a graphical sense, the theory of matroids is pointless. For, in general, the elements of the set S in a matroid can be regarded as a generalization of the set of lines of a graph, whereas there will be no precise analogue of the points of a graph occurring in every matroid. For this reason, we adopt the convention throughout that whenever a subgraph H of a graph G is mentioned below, we will always understand that H is intended as the set of all its lines only, regardless of whether H is a cycle, cocycle, spanning

---

[1]The preparation of this article was supported by a grant from the Office of Naval Research.

tree or forest, subforest, etc.

As a disclaimer, it is not intended that this modest article be construed as a definitive review of the entire field of matroids.

Basic Concepts. If G is a connected graph and T_1 and T_2 are any two spanning trees of G, then clearly for any line e of T_1 there exists a line f of T_2 such that $T_1 - e + f$ is also a spanning tree of G. If G is not connected, the same statement holds for spanning forests.

Similarly, if V is a vector space over a field F and B_1 and B_2 are any two bases of V, then if v is any vector of B_1 there exists a vector u of B_2 such that $B_1 - v + u$ is also a base of V. More generally if U is any subset of the vector space V and B_1' and B_2' are any two maximal sets of independent vectors of U then B_1' and B_2' also have this 'exchange property'.

These are two special cases of a more general theory.

Base Axioms for a Matroid. If S is any finite set, a non-empty family $\mathcal{B}$ of subsets of S is the collection of bases of a matroid on S if
(B1) No member of $\mathcal{B}$ properly contains another.
(B2) If B_1 and B_2 are members of $\mathcal{B}$, then for any $x \in B_1$ there exists $y \in B_2$ such that $B_3 = B_1 - x + y$ also belongs to $\mathcal{B}$.

Theorem 1. If G is any graph and $\mathcal{F}$ denotes the set of spanning forests of G, then $\mathcal{F}$ is the collection of bases of a matroid on the line set $E(G)$.

We call this matroid the cycle matroid of G and will denote it by $M(G)$. The reason for this terminology will be obvious two axiom systems later.

Given any matroid on S, we say that a subset X of S is an independent set if it is contained in some base. We will usually denote the family of independent sets of a matroid by $\mathcal{J}$. Thus a set X of lines of the graph G is independent in its cycle matroid $M(G)$ if and only if X is a subforest of G. Clearly a matroid can always be determined by its independent sets. The next axiom system defines a matroid in terms of such sets.

Independence Axioms for a Matroid. A matroid is a finite set S and a family $\mathcal{J}$ of subsets of S, called independent sets, which satisfy
(I1) $\phi \in \mathcal{J}$.
(I2) If $A \in \mathcal{J}$, then every subset of A is a member of $\mathcal{J}$.
(I3) If $X = \{x_1, \ldots, x_r\} \in \mathcal{J}$ and $Y = \{y_1, \ldots, y_{r+1}\} \in \mathcal{J}$, then there exists $y_i \in Y - X$ such that $X + y_i \in \mathcal{J}$.

For any subset A of S, the rank of A, denoted $r(A)$, is the cardinality of a maximal independent subset of A. It is easy to

verify that the rank function r is well defined and satisfies for
any two subsets A, B ⊂ S,   the 'submodular inequality'

(1) $r(A \cup B) + r(A \cap B) \leq r(A) + r(B)$.

   The <u>rank of the matroid M</u>, written $r(M)$, is the rank of the
set S in M. It is clearly just the cardinality of any base of M.
Thus if G is a graph, the rank of the matroid $M(G)$ is the number
of lines in any maximal spanning forest of G. If X is any subset
of lines of G, then the rank of X in $M(G)$ is the number of lines
in a maximal spanning forest of the subgraph induced by X.

   Let $m(G)$ denote the <u>cycle rank</u> of a graph G. It is well known
that if G has p points, q lines, and k components, then

                         $m(G) = q - p + k$.

Hence the cycle rank of G, and the rank of the matroid $M(G)$, are
related by

(2) $m(G) + r(M(G)) = q$.

   Two matroids M and M' on sets S and S' are <u>isomorphic</u> if
there is a 1-1 map f from S to S' which preserves independence.
Thus if G and H are isomorphic graphs, it is clear that $M(G)$ and
$M(H)$ are isomorphic matroids. However, the cycle matroids of two
nonisomorphic graphs may be isomorphic, as illustrated in Figure 1.

Figure 1. Two trees with 6 lines

Obviously if G and H are any two forests with the same number of
lines, their cycle matroids are isomorphic.

   We say that a subset X of S is <u>dependent</u> in a matroid M if
it is not independent. Thus a set of lines is dependent in $M(G)$ if
and only if it contains a cycle of G. Accordingly we define a subset X to be a <u>circuit</u> of a matroid M if it is a minimal dependent
set. Similarly we say that two elements x and y of S are <u>paral-
lel</u> in M if x,y is a circuit of M, and x is a <u>loop</u> of M if
x is a circuit of M. These definitions have an obvious graphical
origin.

   Whitney [25] postulated an axiom system for a matroid in terms
of its circuits. A more compact system in these terms due to Lehman
[9] is not stated.

<u>Circuit Axioms for a Matroid</u>. A family $\mathcal{C}$ of subsets of S is the
collection of <u>circuits</u> of a <u>matroid</u> if
(C1) No member of $\mathcal{C}$ properly contains another.
(C2) If c_1 and c_2 are members of $\mathcal{C}$ and $x \in c_1 \cap c_2$,   then

$(C_1 \cup C_2) - x$ contains a member of $\mathcal{C}$.

Many other axiom systems for a matroid appear in the literature. We refer in particular to Whitney [25] or Rado [17]. One particularly interesting axiom system uses the conventional notion of closure.

Closure Axioms for a Matroid. A matroid is a set S together with a closure operator c on the subsets of S which satisfies

($\overline{C}$1) If $A \subset B$, then $c(A) \subset c(B)$.

($\overline{C}$2) For any subset A of S, $A \subset c(A) = cc(A)$.

($\overline{C}$3) If $x \in c(A + y)$ and $x \notin c(A)$, then $y \in c(A - x)$.

This last axiom is what Rota [18] calls the 'Steinitz-MacLane exchange axiom'. It is easy to see that for an arbitrary matroid, an element x belongs to c(A) if and only if $r(A + x) = r(A)$. From this it is clear that for a graph G the closure of any set A of lines consists of A together with all lines x lying in some cycle C contained in A + x. For example, let G be the graph of Figure 2.

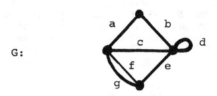

G:

Figure 2. A graph to illustrate closure

Notice first that the loop d belongs to the closure of every set, and that if $f \in A$ then the line g, parallel to f, must belong to c(A). The closure of {a,b} is therefore {a,b,c,d} and in general the closure of any spanning tree is the entire graph G.

We close this section by emphasizing that all of the four axiom systems stated above are equivalent and are just a few examples of the many different systems appearing in the literature. For the most part the proofs of equivalence are routine, though sometimes laborious; see for example Whitney [25].

Duality. The principle of duality is fundamental in matroid theory. Let $\mathcal{C}$* denote the collection of cocycles of a graph G. It is not difficult to verify that $\mathcal{C}$* satisfies the circuit axioms of a matroid. Hence we call this the cocycle matroid of G and denote it by M*(G). Furthermore, it can be shown that the bases of M*(G) are exactly those subsets Y of E(G) such that E(G) - Y is a spanning forest of G. This is just a special case of the following fundamental theorem of Whitney [25].

Theorem 2. If M is a matroid whose bases are B_i, then the collection of sets $S - B_i$ are the bases of a matroid M* on S.

We call M* the dual matroid of M. Clearly the dual of a matroid is unique and the dual of the dual is the original matroid.

Theorem 3. For any graph G, the cycle matroid M(G) and cocycle matroid M*(G) are dual matroids.

If M is any arbitrary matroid we use the following 'co-notation'. If B is a base of M then B is a <u>cobase</u> of M*; C is a <u>cocircuit</u> of M if C is a circuit of M*; and so on. If G is a planar graph and H is its dual graph then the relationship between the matroids of G and H is clear.

Theorem 4. If G is planar and H is the planar graph dual to G, then the cycle and cocycle matroids of G and H are related as follows:

(3)
$$M(G) = M*(H),$$
$$M(H) = M*(G).$$

Example 1. Let G and H be the dual planar graphs of Figure 3. Now {a,b,c} is

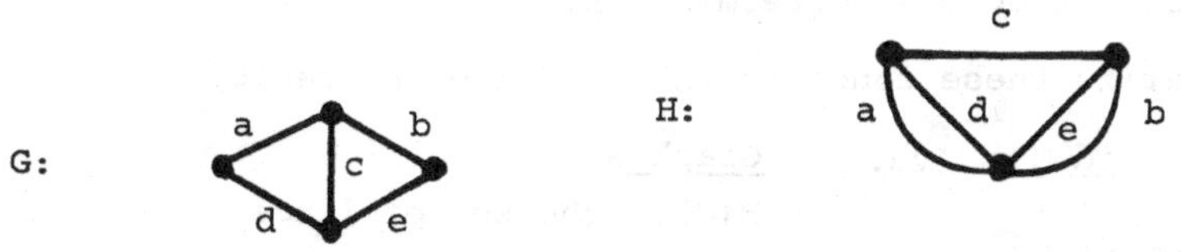

Figure 3. Dual planar graphs

a spanning tree of G, hence {d,e} must be a cobase of M(G) and so a base of M*(G); thus it is a base of M(H) and hence a spanning tree of H. Similarly {a,d} is a cycle of H, hence must be a cocycle of G, and so on.

There exist many matroid theorems of a fairly routine nature which interrelate these concepts.

Theorem 5. A subset B of S is a base of a matroid M if and only if B has non-null intersection with every cocircuit of M and is minimal with respect to this property.

This clearly says, graphically speaking, that T is a spanning forest of a graph G if and only if it is a minimal subset which intersects every cocycle of G. It is impossible to include the details of all such theorems here.

We feel it is worth explicitly stating the basic principle of matroid theory.

Duality Principle. <u>For any theorem about matroids, the dual statement is again a theorem.</u> For example, the dual of Theorem 5 is written Theorem 5*.

Theorem 5*. A set B* of elements is a cobase of M if and only if

B* has non-null intersection with every circuit of M and is minimal with respect to this property.

Specializing to a graph G this asserts that T* is a cotree of a connected graph G if and only if T* has a non-null intersection with every cycle of G and is minimal with respect to this property.

Clearly some theorems are self-dual; for example, it is not difficult to prove an intersection property of circuits and cocircuits.

Theorem 6. For any circuit C and cocircuit C* of a matroid M,

(4) $$|C \cap C^*| \neq 1.$$

The corresponding property in a graph is that the number of lines in the intersection of a cycle and a cocycle is even.

Notice also that in the same way as a matroid can be defined by its bases or circuits, it is also uniquely determined by its cobases or cocircuits and so on. This is because the cobases of M are the bases of M*, hence they determine M* uniquely, and by the uniqueness of the dual they must determine M.

We summarize these concepts in the following table:

Matroid M vs.	Graph G
set S	E(G), the set of lines
element	line
base	maximal spanning forest
independent set	subforest
circuit	cycle
rank r(M)	cocycle rank m*(G)
cobase	complement of maximal spanning forest
cocircuit	cocycle
corank r*(M)	cycle rank m(G)

From Graphs to Matroids. In this section we give examples of the way in which graphical theorems can be generalized to obtain results for matroids.

Example 2. If T is a spanning tree of the graph G and x is a line of G - T, then it is well known that there is a unique cycle C of G such that $x \in C \subset T + x$.

It is easy to prove that the corresponding statement holds for matroids.

Theorem 7. If B is a base of the matroid M and x is an element of S - B, then there is a unique circuit C of M such that $x \in C \subset B + x$.

The circuit C is called the fundamental circuit of x in the base B. By the duality principle we obtain the next result immediately.

Theorem 7*. If B* is a cobase of the matroid M and x is an element of S - B*, then there is a unique cocircuit C* of M such that $x \in C^* \subset B^* + x$.

This is of course the matroid result generalizing the dual of Example 2.

<u>Example 3</u>. If T^* is a cotree of G and x is a line of G not in T^* , then there is a unique cocyle C^* of G such that $x \in C^* \subset T^* + x$.

<u>Example 4</u>. Let G - A denote the graph obtained from G by deleting the lines A and let $G|A$ denote the graph obtained by contracting the lines of A. It is clear that the cycles of G - A are just those cycles of G which are contained in $E(G) - A$. Tutte [19] generalized these ideas to matroids.

If M is any matroid and $T \subset S$, define the <u>reduction minor</u> M x T to be the matroid on T which has as its circuits just those circuits of M which are contained in T. Similarly, let the <u>con-traction minor</u> M·T be that matroid whose cocircuits are just those cocircuits of M which are contained in T. A useful algebra of con-traction and reduction is developed in Tutte [19].

<u>Theorem 8</u>. For any matroid M and any subset T of S,

(5) $(M \times T)^* = (M^* \cdot T)$.

By a simple duality argument this implies

(5*) $(M \cdot T)^* = (M^* \times T)$.

As an illustration of this theorem, take G to be a planar graph and let H be its dual graph. Now let $G' = G - A$ and H' be the graph obtained from H by contracting the lines of H corresponding to A. We see that the cycle matroids of G' and H' are dual. In Figure 4, let $A = \{a,c\}$ in G, the same graph chosen at random as used in Figure 3.

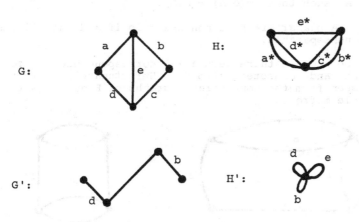

<u>Figure 4</u>. <u>A construction for obtaining dual matroids</u>

Matroid Connection. It is clear that there is no concept in a general matroid corresponding to the notion of a point of a graph. Thus connection in the usual graphical sense has no matroid counterpart. However, 2-connectedness in graph theory extends naturally to matroids. Recall that a graph G with at least three points is 2-connected if and only if it is a block and thus if and only if every pair of distinct lines are contained in a cycle. Whitney [25] says that a matroid is connected or nonseparable if there exists no proper subset A of S such that

(6) $r(A) + r(S-A) = r(S)$.

It turns out that this notion of connection is equivalent to the following. Let ~ be a binary relation on S defined by $x \sim y$ if either $x = y$ or there exists a circuit C of M containing both x and y. It is easy to prove that ~ is an equivalence relation and that M is connected in the sense of (6) if and only if S is the single equivalence class under ~ .

Lemma 9.1. If x and y are distinct members of a circuit C of M then there is a cocircuit C* of M containing x and y and no other members of C.

Theorem 9. A matroid M is connected if and only if its dual M* is connected.

In terms of graph theory, it is clear that G has a connected cycle matroid if and only if G is nonseparable.

Corollary 9.1. A graph is a block if and only if every pair of distinct lines are contained in a cocycle.

Corollary 9.2. The cycle matroid of G is connected if and only if the cocycle matroid of G is connected.

Corollary 9.3. Graph G is nonseparable if and only if there is no set of lines A such that $m(G-A) = m(G)$.

Corollary 9.4. A planar graph is nonseparable if and only if its dual graph is nonseparable.

It would be nice if there were a 1-1 correspondence between nonseparable graphs and connected matroids. Unfortunately, this is not so as may be seen from the nonseparable graphs of Figure 5 which have isomorphic cycle matroids.

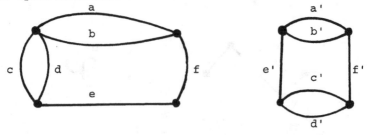

Figure 5.   Nonseparable graphs which have
             isomorphic cycle matroids.

Whitney [26] calls such graphs "2-isomorphic."

Binary and Graphic Matroids. A matroid M is called graphic if there exists some graph G such that M is isomorphic to the cycle matroid of G. Not all matroids are graphic; the smallest non-graphic matroid is the following matroid M_0 defined by $S_0 = \{1, 2, 3, 4\}$ and with its independent sets, all subsets of at most 2 elements.

A matroid M is cographic if there exists a graph G such that M is isomorphic to the cocycle matroid of G. It is easy to see that M is therefore cographic if and only if its dual M* is graphic, and that a matroid is both graphic and cographic if and only if it is isomorphic to the cycle matroid of some planar graph G. Notice that the self-dual matroid M_0 is also not cographic. Thus a matroid M is graphic but not cographic if and only if M is isomorphic to the cycle matroid of a nonplanar graph. Hence the smallest graphic but not cographic matroids are the cycle matroids of K_5 and K(3,3). And by duality, the smallest cographic but not graphic matroids are the cocycle matroids of K_5 and K(3,3).

A first necessary condition to place on a matroid in order for it to be graphic or cographic is immediate.

For any collection C_1, C_2, ..., C_k of circuits of M, the symmetric difference $C_1 \oplus C_2 \oplus \dots \oplus C_k$ is the union of disjoint circuits of M.

Matroids which have this property are called binary matroids. (This definition can be shown to be equivalent to that given by Tutte [19]in terms of chain groups.) The next theorem characterizes such matroids.

Theorem 10. The following conditions on a matroid are equivalent.
(a) M is binary.
(b) For any circuit C and cocircuit C* of M, the cardinality of $C \cap C^*$ is even.
(c) For any base B and any circuit C, if $x_1, x_2, \dots, x_k$ are the elements of C - B, and C_i is the fundamental circuit of x_i in B, then $C = C_1 \oplus C_2 \oplus \dots \oplus C_k$.

From the symmetric nature of (b) above, we immediately have the next result.

Corollary 10.1. A matroid is binary if and only if its dual is binary.

The next two results are straightforward consequences of the definition of a binary matroid.

Theorem 11. If M is binary and $T \subset S$, then both the contraction minor M·T and the reduction minor M × T are binary.

Theorem 12. For any graph G, both its cycle matroid M(G) and its cocycle matroid M*(G) are binary.

Not all binary matroids are either graphic or cographic.

Example 5. Let $S = \{1, 2, 3, 4, 5, 6, 7\}$ and let M have as bases all 3-subsets of S except

$$\{1,2,6\} \quad \{1,4,7\}, \quad \{1,3,5\}, \quad \{2,3,4\}$$
$$\{2,5,7\} \quad \{3,6,7\}, \quad \{4,5,6\} \ .$$

This matroid is often described as the <u>Fano matroid</u> as it is easily seen to be the matroid obtained by taking the familiar Fano config-uration in which a set of 3 points is called independent if these points are not colinear. The matroid dual of the Fano matroid has rank 4, has been called the "heptahedron matroid", and is also binary but neither graphic nor cographic.

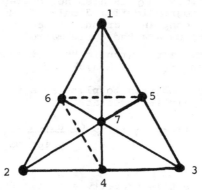

Figure 6. The Fano Configuration

Nevertheless, several graphical theorems can often be regarded as special cases of theorems for binary matroids, but not for ma-troids in general. We illustrate this with some simple examples.

Example 6. Connected graphs in which every line belongs to at most one cycle were originally described by Harary and Uhlenbeck [6] as 'Husimi trees'. They are more aptly described in [5] as <u>cacti</u> (especially since they are not always trees). It is easy to show that a connected graph with every cycle of odd length is a cactus. Accordingly, we say that a matroid is a <u>Husimi matroid</u> if every ele-ment belongs to at most one circuit and a matroid is <u>odd</u> if all of its circuits have odd cardinality. Manvel has shown that if M is binary and odd then it is a Husimi matroid. This is not true for non-binary matroids, as shown by the matroid on a set S of 5 elements which has as its circuits all the 3-subsets of S.

Example 7. We call M a <u>bipartite matroid</u> if every circuit of M is even (has even cardinality). If there exists a collection of dis-joint circuits C_i of M such that

$$S = C_1 \cup C_2 \cup \ldots \cup C_k \ ,$$

then we say that M is an eulerian matroid. It is not difficult to prove that a binary matroid M is eulerian if and only if M* is bipartite. Again, this is not true for non-binary matroids; we refer to Welsh [22] for details.

Example 8. Incidence Matrices. If C_i, $i = 1, \ldots, m$, are the
cycles of a graph G with q lines, the cycle matrix of G is the
$m \times q$ matrix $C(G) = [c_{ij}]$ in which $c_{ij} = 1$ if line e_j is a
member of the cycle C_i and is 0 otherwise. The cocycle matrix
$C^*(G)$ is defined analogously. It is well known that $C(G)$ has rank
equal to $m(G)$, the cycle rank of G, and hence by (2) is equal to
the rank of the cocycle matroid $M^*(G)$.

These may be generalized to binary matroids as follows. Let C_i,
$i = 1, \ldots, m$ be the circuits of a matroid M on a set S of cardinality n. Let $C(M)$ be the $m \times n$ matrix whose i,j entry is 1 if
the element x_j of S is a member of the circuit C_i and is 0
otherwise. Then $C(M)$ is called the circuit matrix of M and the
cocircuit matrix $C^*(M)$ is defined analogously.

Theorem 13. If M is binary, then the rank of the circuit matrix
$C(M)$ is $r(M^*)$ and the rank of the cocircuit matrix $C^*(M)$ is $r(M)$.

Another property of the cycle and cocycle matrices of a graph G
is that they are orientable in the sense that it is possible to assign
negative signs to some of the non-zero entries of $C(G)$ and $C^*(G)$
so that the matrices $C_0(G)$ and $C_0^*(G)$ with 0, +1, and -1 entries so obtained have the property that each row of $C_0(G)$ is a vector orthogonal to each row of $C_0^*(G)$. We know that such an assigna-
tion is far from being unique, since one can be found by just taking
D to be any digraph obtained by orienting the lines and cycles of G
and then assigning signs to the entries of $C(G)$ and $C^*(G)$ in the
natural way. (If the orientation of line i agrees with the orientation assigned to cycle C_j of G, then the i,j entry is +1;
otherwise it is -1.)

We use this property of G to define a matroid M to be orientable if it is possible to assign positive and negative signs to the
non-zero entries of the circuit and cocircuit matrices $C(M)$ and
$C^*(M)$ in such a way that if $C_0(M)$ and $C_0^*(M)$ denote the matrices
with these orientations, then $C_0(M)$ and $C_0^*(M)^T$ are orthogonal
matrices.

Theorem 14. Every graphic matroid is orientable. By duality, every
cographic matroid is also orientable.

However, not every binary matroid is orientable. For example,
the Fano matroid is non-orientable. A minor of a matroid M is any
matroid M' on a subset T of S obtained by a succession of reductions or contractions of M. Tutte [19] obtained the following
characterization of orientable matroids, using minors.

Theorem 15. A matroid is orientable if and only if it is binary and
contains no minor isomorphic either to the Fano matroid or its dual.

In fact, Tutte [19] actually proved that this condition characterizes what he called "regular matroids," but Minty [11] showed that

M is regular in the sense of Tutte if and only if it is orientable.

Now we state the fundamental result of Tutte [19], which solved an outstanding problem in the electrical engineering literature!

Theorem 16. A matroid M is graphic if and only if it is orientable and has no minor which is isomorphic to the cocycle matroid of K_5 or K(3,3).

By duality we have a criterion for cographic matroids.

Theorem 16*. A matroid M is cographic if and only if it is orientable and has no minor isomorphic to the cycle matroid of either K_5 or K(3,3).

In view of the analogy between this and Kuratowski's theorem giving conditions for a graph to be planar, it is not surprising that there is a matroid analogue of MacLane's theorem [10] on planar graphs. We say that a family $C_1^*, \ldots, C_{r(S)}^*$ of cocircuits of M form a 2-complete basis of the vector space generated by the cocircuits of M with respect to the symmetric difference operator if they are a basis in the usual sense and also no element of S is a member of more than two of the C_i^*. Then we have the following theorem of Welsh [23].

Theorem 17. A matroid M is graphic if and only if it has a 2-complete basis of cocircuits.

Theorem 17*. A matroid M is cographic if and only if it has a 2-complete basis of circuits.

Application of Matroid Theory to Graphs. Apart from its intrinsic purely mathematical interest and the rather elegant manner in which it can handle cocycles of graphs, matroid theory is a very powerful tool in many of the combinatorial problems connected with graphs. We will now point out one of these applications by way of illustration.

One of the most fundamental theorems in this area of combinatorics is the following matroid generalization of Hall's theorem which was proved by Rado [15] in 1942 and passed almost unnoticed for two decades. If $\mathcal{A} = \{A_1, A_2, \ldots, A_n\}$ is any family of subsets of the finite set S, we say that $X = \{x_1, x_2, \ldots, x_r\}$ is a partial transversal of $\mathcal{A}$ of length k, if there exist distinct members $A_{i_1}, A_{i_2}, \ldots, A_{i_k}$ of $\mathcal{A}$ such that $x_j \in A_{i_j}$, $1 \le j \le k$. A transversal of $\mathcal{A}$ is a partial transversal of length n. We may now state Rado's theorem in the following form.

Theorem 18. If M is a matroid on S and $\mathcal{A} = \{A_1, \ldots, A_n\}$ is any family of subsets of set S, then $\mathcal{A}$ has a transversal X which is independent in M if and only if for any subset $J \subset \{1, \ldots, n\}$, writing $A(J) = \bigcup_{i \in J} A_i$, we have

$$r(A(J)) \ge |J|.$$

The importance of this theorem is that merely by constructing 'useful matrods' on S, we can obtain many of the Hall-type theorems proved by Hoffman-Kuhn [7], [8] and others by <u>ad hoc</u> or linear programming methods. For example, if we let M be the trivial matroid in which every subset is independent, then we get Hall's theorem [2]. By taking M to have as bases only those subsets of S having cardinality n and containing a prescribed subset E, we get the conditions for Q to have a transversal containing the prescribed subset E. Many other applications exist, see for example Mirsky and Perfect [12], or Welsh [25].

<u>Arboricity</u>. We now turn to a deep theorem of Nash-Williams [13] on matroids, which is particularly interesting to graph theorists and is an example of the use of mathematical generalization. Tutte [20] and Nash-Williams [14] answer the following two problems by ingenious but intricate graph theoretic arguments:
(P1) When does a graph G have k line-disjoint spanning trees?
(P2) What is the minimum number of disjoint subforests whose union is G?
The answer to (P2) is called the <u>arboricity</u> of G; Beineke [1] studied this problem constructively for complete graphs and bigraphs.

These are special cases of the more general questions:
(P1') When does a matroid M have k disjoint bases?
(P2') If M is a matroid on S, when is S the union of k independent sets?

Lehman [9] first noticed that the graphical problems (P1) and (P2) are related via matroid theory to the "Shannon switching game" and Edmonds [3] and [4] answered problems (P1') and (P2'). We now show that they can be answered quite easily using the following matroid theorem of Nash-Williams [13].

Let M_1, M_2, ..., M_k be matroids on the same set S. Let their respective ranks be r_i, and let J_i be the family of independent sets of M_i. Let J be the family of subsets of S of the form $X_1 \cup X_2 \cup \ldots \cup X_k$ where $X_i \in J_i$.

<u>Theorem 19</u>. The family J of subsets of S are the independent sets of a matroid on S, denoted by $M_1 \vee M_2 \vee \ldots \vee M_k$, which has rank r given by

(7) $$r = \min_{A \subseteq S} [r_1(A) + \ldots + r_k(A) + |S-A|].$$

We show how to use this result to answer (P1') and (P2'). Given a matroid M, take $M_i = M$ for all i = 1, ..., k. Then by Theorem 20, the matroid M has k disjoint bases if and only if the matroid $M(k) = M \vee \ldots \vee M$ (k times) has rank equal to $kr(M)$.

<u>Corollary 19.1</u>. Matroid M has k disjoint bases if and only if for any subset A of S,

$$|S-A| \geq k[r(M) - r(A)] .$$

Similarly, S is the union of k independent sets if and only

if M(k) has rank equal to |S| .

Corollary 19.2. If M is a matroid on S, then S is the union of k independent sets if and only if for all A ⊂ S,

$$kr(A) + |S-A| \geq |S| .$$

Applying this to a graph G, we get a condition which determines its arboricity.

Theorem 20. The arboricity of a graph G is k if and only if for any set A of lines of G,

$$km*(A) \geq |A| .$$

Conclusion. We hope that the reader is now in a position to appreciate the important link between graph theory and matroids. This link is forged in two distinct ways:
  (a) Known theorems in graph theory can be generalized to
        give theorems for matroids and hence new theorems about
        transversals and vector spaces.
  (b) By considering the more general matroid structure, one
        often gets a clearer idea of the nature of the problem
        and particularly by using duality obtains simpler proofs
        of graph theoretical results.

As an example of a problem of type (a) we leave the reader with the following conjecture. From Dirac's well known graphical theorem, it is clear that if G is a block in which every cocycle has cardinality not less than p/2, then G is hamiltonian. This prompts the following question about matroids:

Conjecture. If M is a connected binary matroid in which every co-circuit has cardinality not less than (r(M) + 1)/2, then M has a circuit of cardinality r(M) + 1.

REFERENCES

1. L.W. Beineke, Decompositions of complete graphs into forests, Magyar Tud. Akad. Mat. Kutato Int. Közl. 9 (1964), 589-594.

2. P. Hall, On representatives of subsets, J. London Math. Soc. 10 (1935), 26-30.

3. J. Edmonds, Minimum partition of a matroid into independent subsets, J. Res. Nat. Bur, Stand. 69B (1965), 67-72.

4. J. Edmonds, Lehman's switching game and a theorem of Tutte and Nash-Williams, J. Res. Nat. Bur. Stand. 69B (1965), 73-77.

5. F. Harary, Graph Theory, Addison-Wesley, Reading, 1969.

6. F. Harary and G.E. Uhlenbeck, On the number of Husimi trees, I. Proc. Nat. Acad. Sci., USA 39 (1953), 315-322.

7. A.J. Hoffman and H.W. Kuhn, On systems of distinct representatives, Linear Inequalities and Related Systems (Annals of Math.

Stud. 38, Princeton), (1956), 199-206.

8. A.J. Hoffman and H.W. Kuhn, Systems of distinct representatives and linear programming, Amer. Math. Monthly 63 (1956), 455-460.

9. A. Lehman, A solution of the Shannon switching game, J. Soc. Indust. Appl. Math. 12 (1964), 687-725.

10. S. MacLane, A combinatorial condition for planar graphs, Fund. Math. 28 (1937), 22-32.

11. G.J. Minty, On the axiomatic foundations of the theories of directed linear graphs, electrical networks and network programming, J. Math. Mech. 15 (1966), 485-520.

12. L. Mirsky and H. Perfect, Applications of the notion of independence to problems of combinatorial analysis, J. Combinatorial Theory 2 (1967), 327-357.

13. C.St.J.A. Nash-Williams, On applications of matroids to graph theory, Theory of Graphs International Symposium Rome, Dunod, 1968, 263-265.

14. C.St.J.A. Nash-Williams, Edge disjoint spanning trees of finite graphs, J. London Math. Soc. 36 (1961), 445-450.

15. R. Rado, A theorem on independence relations, Quart. J. Math. Oxford 13 (1942), 83-89.

16. R. Rado, Note on independence functions, Proc. Lond. Math. Soc. 7 (1957), 300-320.

17. R. Rado, Abstract linear dependence, Colloq. Math. 14 (1966), 257-264.

18. G.C. Rota, On the foundations of combinatorial theory, I: Theory of Möbius functions, Zeit. Wahrschein. 2 (1964), 340-368.

19. W.T. Tutte, Lectures on matroids, J. Res. Nat. Bur. Stand. 69B (1965), 1-48.

20. W.T. Tutte, On the problem of decomposing a graph into n-connected factors, J. Lond. Math. Soc. 36 (1961), 221-230.

21. D.J.A. Welsh, Matroids and their Applications, Seminar Notes, Univ. of Michigan (to appear).

22. D.J.A. Welsh, Euler and bipartite matroids, J. Combinatorial Theory (to appear).

23. D.J.A. Welsh, On the hyperplanes of a matroid, Proc. Camb. Phil. Soc. (to appear).

24. D.J.A. Welsh, Applications of a theorem by Rado, Mathematika (to appear).

25. H. Whitney, On the abstract properties of linear dependence, Amer. J. Math. 57 (1935), 509-533.

26. H. Whitney, 2-isomorphic graphs, <u>Amer. J. Math</u>. 55 (1933),
     245-254.

ON CLASSES OF GRAPHS
DEFINED BY
SPECIAL CUTSETS OF LINES[1]

Stephen Hedetniemi, University of Iowa

Abstract. In this paper we present a new method for studying
graphs. Generally speaking this involves decomposing a graph into
two disjoint subgraphs which are connected by special sets of lines.
We consider four types of connections between these subgraphs, i.e.,
those for which the set of connecting lines describes a function, a
homomorphism, a permutation, or an automorphism.

We consider this manner of decomposing a graph to be useful for
studying a wide variety of parameters and properties of graphs. To
illustrate this we obtain results relating to such concepts as arboricity, thickness, biparticity, and chromatic number. We derive a
method for constructing new classes of critical graphs and obtain
several isomorphism theorems for classes of permutation graphs, one
of which involves the group theoretic concept of a double coset.

1. Introduction. Suppose we are given a graph $G = (V, E)$ and we
partition the set of points of G into two nonempty sets V_1 and
V_2. Then G can be expressed as the union of two induced subgraphs
$\langle V_1 \rangle$ and $\langle V_2 \rangle$ together with the set of lines E_{12} between points
of V_1 and V_2. This set of lines, E_{12}, can be viewed as defining
a (symmetric) binary relation, say π, on the set $V_1 \times V_2$, and can
also be considered as defining a bipartite subgraph, the removal of
which separates (disconnects) G.

Conversely, given two graphs G and H and a (symmetric) bi-
nary relation $\pi \subseteq V(G) \times V(H)$, we may define a π-graph, $G\pi H$,
where $V(G\pi H) = V(G) \cup V(H)$ and $E(G\pi H) = E(G) \cup E(H) \cup \pi$.

It is of interest first of all to list various special kinds of
binary relations π which can be used to separate (or connect)
graphs in this way. Let G and H be two disjoint graphs and let
$\pi \subseteq V(G) \times V(H)$, then
   if $\pi = \emptyset$ (empty relation), then $G\pi H$ is the disjoint union of
G and H;
   if $\pi = \{(u,b) \mid v \in V(G), v \in V(H)\}$, then $G\pi H$ is the join, usually denoted $G + H$, of G and H;
   if π is a function, say f, from $V(G)$ to $V(H)$, then we
will say that GfH is a function graph;
   if π defines a homomorphism, say φ, from G to H, then
$G\varphi H$ is a homomorphism graph;
   if $G = H$ and π is a 1-1 function from $V(G)$ onto $V(H)$,

[1]Research supported in part by the Office of Naval Research, Contract
Number NR 043-367.

i.e., π can be considered to be a permutation α of $V(G)$, then $G\alpha H$, or $G\alpha G$, is a <u>permutation graph</u> (c f. Chartrand and Harary [4], who used the notation $P_\alpha(G)$);

finally, if $G = H$ and π is an automorphism α of G, then $G\alpha G$ is an <u>automorphism graph</u>.

It is the purpose of this paper to explore these classifications of graph, to consider which graphs fall into these classifications, and to discover properties held by members of these classes. In particular we obtain bounds for the chromatic numbers of certain classes of graphs $G\pi H$ and we derive, somewhat unexpectedly, several results in other areas of graph theory, such as arboricity, thickness, biparticity and the construction of critical graphs, which are suggested by results on π-graphs.

2. <u>Function graphs</u>. According to our description of function graphs given earlier, <u>every</u> graph $G = (V,E)$ can be expressed as a function graph, i.e. let $u \in V$ and set $V_1 = V - \{u\}$, $V_2 = \{u\}$. Then $G = \langle V_1 \rangle \cup \langle V_2 \rangle \cup E_{12}$, where the set of lines E_{12} defines a function from a subset V_u of V_1 onto $V_2 = \{u\}$; V_u is simply the set of points adjacent to u in G. In other words, every graph can trivially be separated by a (partial) function (unless possibly $G = K_n$). Henceforth we shall say that a graph G is a <u>function graph</u> if there exists a partition V_1, V_2 of $V(G)$ such that $G = \langle V_1 \rangle \cup \langle V_2 \rangle \cup E_{12}$, where E_{12} defines a <u>function</u> from <u>all</u> the points of V_1 to V_2. Furthermore, if G is decomposed into a function graph as above, we will write $G = G_1 f G_2$, where $G_1 = \langle V_1 \rangle$, $G_2 = \langle V_2 \rangle$ and $f = E_{12}$; in such a decomposition of G, V_1 is the <u>domain</u>, V_2 is the <u>range</u> of G, and we will say that G is <u>functional</u>. If a graph G cannot be expressed as a function graph then G is <u>afunctional</u>.

Figure 1 illustrates a function graph; G is the well-known cubic graph on 10 points containing a bridge. Two representations of G as a function graph are indicated; in Figure 1a the four encircled points define a <u>range</u> for G, i.e., every other point is adjacent to exactly one of these four points; and in Figure 1b the four encircled points define a <u>domain</u> for G.

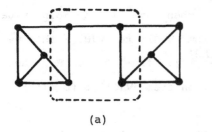 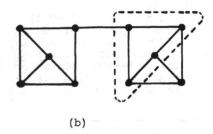

(a)   (b)

Figure 1

The question is now raised anew: which graphs are function graphs?

One can readily assert that the following classes of graphs are function graphs:
   (i) any graph with a point u of degree 1;   the domain of such a graph can consist only of the point u;   consequently, any <u>tree</u> or <u>forest</u> is a function graph;
   (ii) any graph with two adjacent points u,v both of degree 2; the domain of such a graph can consist only of the points u and v; this class of function graphs includes all <u>cycles</u>;
   (iii) any graph G containing one point u which is adjacent to all other points of G,   i.e., graph which can be expressed in the form $G = K_1 + H$;   the range of such a graph can consist of only the point u;   this class includes all the <u>complete graphs</u> K_n;
   (iv) any <u>complete</u> <u>bipartite</u> <u>graph</u> $K_{m,n}$;   the range of such a graph can consist of any two adjacent points;
   (v) any bipartite graph G containing a path u_1, u_2, u_3 of length two, the points of which have degrees 2,3, and 2, respectively; these three points can constitute the domain of G.
   (vi) to generalize (ii) above, any graph containing n mutually adjacent points each of degree n;   these n points can define a domain for G;   this class includes any regular graph of degree n containing a complete subgraph on n points;
   (vii) every cubic graph is a function graph; such a graph must first of all contain a cycle, therefore the domain for a cubic graph G can consist of the set of points in any smallest cycle of G.

There exist graphs which cannot be expressed as function graphs, an example is the graph K(2,2,2).   The smallest afunctional graphs known to date are given in Figure 2.

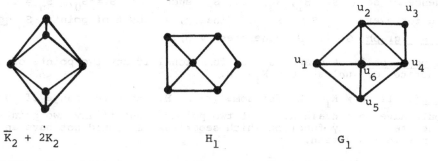

$\overline{K}_2 + 2K_2$ H_1 G_1

Figure 2

Because of (i) and (iv) above, one might think that perhaps all bipartite graphs are functional; however, this is not the case. Figure 3 illustrates an afunctional bipartite graph.

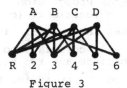

Figure 3

Verification that a given graph G is afunctional, is, almost by definition, a tedious matter, as it can involve checking all the possible partitions V_1, V_2 of $V(G)$. In certain cases, however, such verification is greatly simplified by considering what is necessary, say, if a given subset of points were to lie in the range of the supposed function. For example, consider the graph G_1 of Figure 2.

Since $G_1 \neq K_1 + H$ for some graph H, if G_1 were a function graph for some function f, then at least two points would have to lie in the range of f. Suppose, for example, u_1 and u_3 were in the range. Then since u_2 is adjacent to both u_1 and u_3, u_2 could not lie in the domain of f, and therefore must also lie in the range of f. But now, since u_6 is adjacent to u_1 and u_2, both of which lie in the range, u_6 must also lie in the range. Furthermore, if u_6 must lie in the range of f, so also must u_4 and u_5. Consequently, we see that if G were a function graph, then points u_1 and u_3 could not both lie in the range. Arguments such as these lead us to the following determination of several classes of afunctional graphs.

Let us say that two points u,v _directly generate_ a third point w if both u and v are adjacent to w. Similarly, a set of points S _directly generates_ a point w if either $w \in S$ or there are two points in S adjacent to w. A set of points S _directly generates_ a set T, written $S \Rightarrow T$, if every point of T is directly generated by S. A set S _generates_ a set T if there exists a finite sequence of sets S_0, S_1, S_2, ..., S_n such that $S = S_0$, $S_n = T$, and for every i, $S_i \Rightarrow S_{i+1}$. Finally, a subset of points S _generates a graph_ G if S generates $V(G)$.

Proposition 1. A graph G is afunctional if any two points of G generate G, unless $G = K_1 + H$.

Proof. If $G \neq K_1 + H$ for some graph H, then the range of G would have to contain at least two points. But if any two points generate G, any function which separates G could not have any points in its domain.

Both of the graphs $\overline{K}_2 + 2K_2$ and G_1 in Figure 2 are examples of Proposition 1; in fact, all graphs which are afunctional because of Proposition 1 must be blocks of diameter 2. Notice that the graph H_1 of Figure 2 is a subgraph of G_1 which is afunctional but does not satisfy the conditions of Proposition 1. Furthermore, if $G_1 \geq G \geq H_2$, then if G is functional (or afunctional) one cannot in general infer that either H_1 or H_2 is functional (or afunctional). It also follows as a corollary to Proposition 1 that if a graph G is afunctional because any two points of G generate G_1 and if G' is obtained from G by adding a line to G so that

$G' \neq K_1 + H_1$ then G' will also be afunctional.

The following extension of Proposition 1 provides us with more afunctional graphs.

<u>Proposition 2</u>. A graph $G \neq K_1 + H$ is afunctional if any two adjacent points generate G and every maximal independent set of points S of G directly generates a point not in S.

<u>Proof</u>. If G were functional, say $G \cong G_1 f G_2$, then the range of f could not contain any adjacent pair of points, since by assumption these two points would generate G. The range could not consist of an independent set of points which was not maximal, for then there would exist a point in the domain not adjacent to any point of the range, and hence f would not be totally defined. Finally, we see that the range could not consist of a maximal independent set of points S, for by assumption there would be a point not in S, therefore in the domain of f, which was adjacent to two points in S.

The graph in Figure 4 illustrates Proposition 2.

Figure 4

It can be seen, from Proposition 1, that all of the complete m-partite graphs $K_{n_1, n_2, \ldots, n_m}$ are afunctional when $m \geq 3$ and $n_i \geq 2$ for all i. But, this fact also follows from the following result.

<u>Proposition 3</u>. The join of two graphs $G+H$ is a function graph if and only if either $G+H \cong K_1+L$ or both G and H contain isolated points.

<u>Proof</u>. The sufficiency is obvious. Suppose $G+H$ is a function graph; and suppose further that $G+H \neq K_1+L$, for some graph L.
Then the range of $G+H$ must contain at least two points, say u,v.
We consider three cases:
    (i) $u,v \in V(H)$; then no point of G can be in the domain since each point of G is adjacent to both u and v, and consequently the range of $G+H$ must contain at least $V(G) \cup \{u,v\}$. But now, no remaining point of H can lie in the domain, since each of these points is adjacent to every point of G.
    (ii) $u,v \in V(G)$; the argument here is the same as in (i).
    (iii) $u \in V(G)$, $v \in V(H)$; in this case, any point in the domain, say $u' \in V(G)$ cannot be adjacent to point u; and similarly if $v' \in V(H)$ is in the domain then v' cannot be adjacent to v.
Finally no other point of either G or H could lie in the range,

because then the arguments in case (i) and (ii) above would apply.
Thus if G+H is a function graph, and G+H ≠ K₁+L, then the range
must consist of two points u,v, such that u∈V(G), v∈V(H); all
other points of G+H must lie in the domain, no point u'∈V(G) - {u}
can be adjacent to u, and no point v'∈V(H) - {v} can be adjacent
to v. Thus, both G and H contain an isolated point, u and v,
respectively.

Thus we see from Proposition 3 that for most graphs G,H the
join operation produces afunctional graphs G+H.

<u>Theorem 4</u>. If every block of G is afunctional, then G is afunc-
tional.

<u>Proof</u>. We proceed by induction on the number n of blocks of G.
Clearly, if n = 1 the assertion is trivial. Assume then that every
graph consisting of n - 1 afunctional blocks is afunctional. Let
G consist of n afunctional blocks, and consider the block-cutpoint
tree of G, bc(G) (cf. Harary and Prins [7]). We assume here without loss of generality that G is connected. Let B denote an end-
block (endpoint) of bc(G) and let w be the corresponding cutpoint
of B. Assume that G is functional and let D,R ⊆ V(G) be a domain and range, respectively, for G.

We consider now V(B) ∩ D and V(B) ∩ R, i.e., how do D and
R partition V(B)?

Case 1: If V(B) ∩ D = ∅, i.e., V(B) ⊆ R, then by removing
V(B) - w from G we would obtain a graph G' having n - 1 afunc-
tional blocks which was functional, having domain D and range
R - {V(B)} ∪ {w}; but this contradicts our induction hypothesis.

Case 2: V(B) ∩ D = ∅, i.e., V(B) ⊆ D. This is impossible
since every point of V(B) would have to be adjacent with one point
of R, since the points of V(B), with the exception of w, are
not adjacent with any point not in B, and since |V(B)| ≥ 2.

Case 3: D_B = V(B) ∩ D ≠ ∅, R_B = V(B) ∩ R ≠ ∅. Since by assumption B is afunctional, the decomposition D_B,R_B of V(B) cannot
be functional, i.e., either (a) one point of D_B is adjacent to two
points of R_B, or (b) one point of D_B is not adjacent with any
point of R_B. But case (a) cannot occur, otherwise D,R would not
provide a functional representation of G. In case (b) the point
would have to be adjacent with exactly one point not in V(B) but in
R - R_B. Thus this point would have to be w. Consequently, if we
remove from G the points V(B) - w we will obtain a functional representation of a graph consisting of n - 1 afunctional blocks, a
contradiction.

Before moving away from the problem of characterizing function
graphs, we would like to mention that so far we do not have a reasonable explanation as to why the graph H_1 in Figure 2 is afunctional.

The following results provide us with additional insights into the
nature of function graphs.

Proposition 5. Every function graph G+H is isomorphic to a function graph G' f H' where G' is a connected graph.

Proposition 6. Let H be a connected graph; then
    (i) if GfH is planar, then G is outerplanar,
    (ii) if GfH is outerplanar, then G is acyclic, and
    (iii) if GfH is acyclic, then G is totally disconnected.

Proof. We will prove only part (i) [cf. [3] for further discussion of planar, outerplanar, acyclic, and totally disconnected graphs]. Suppose G is not outerplanar, i.e., G contains a subgraph which is homeomorphic to either $K_{2,3}$ or K_4. We have assumed that H is connected; thus let us contract H to a single point, i.e., we can contract GfH to $\dot{G}+K_1$. But if G contains a subgraph homeomorphic to either $K_{2,3}$ or K_4, then clearly $G+K_1$ contains a subgraph homeomorphic to either $K_{3,3}$ or K_5. This means that $G+K_1$ is not planar; and thus GfH is not planar, since we obtained $G+K_1$ from GfH by a contraction. Since GfH was assumed planar, we have a contradiction. Thus G must be outerplanar.

A closer look at the proof of Proposition 6 reveals that a more general statement can be made as follows; we omit the proof.

Proposition 6a. Let H(G) be a connected graph and let $\pi \subseteq V(G) \times V(H)$ be a symmetric binary relation whose domain (range) is all of V(G) (V(H)), then
    (i) if GπH is planar, then G(H) is outerplanar,
    (ii) if GπH is outerplanar, then G(H) is acyclic,
    (iii) if GπH is acyclic, then G(H) is totally disconnected.

3. **Thickness and Arboricity.** The **thickness** t(G) of a graph G is the minimum number of line disjoint planar subgraphs whose union equals G.

Proposition 7. Let GfH be a partial function graph, i.e., f is a partially defined function from V(G) to V(H), then
    (i) t(G) < t(H) implies t(GfH) = t(H); and
    (ii) t(G) ≥ t(H) implies t(GfH) ≤ t(G) + 1.

Proof. Let t(G) = m and let G_1, G_2, ..., G_m be m line-disjoint planar subgraphs whose union is G; similarly, let t(H) = n > m, and let H_1, H_2, ..., H_n be the corresponding planar subgraphs of H. Then clearly, $H_n \cup f$ is planar. Thus, $H_1 \cup G_1$, $H_2 \cup G_2$, ..., $H_m \cup G_m$, H_{m+1}, ..., $H_n \cup f$ are n line disjoint planar subgraphs whose union is GfH. A similar argument in case m ≥ n shows that $G_1 \cup H_1$, $G_2 \cup H_2$, ..., $G_n \cup H_n$, G_{n+1}, ..., G_m, f constitutes a set of t(G)+1 line disjoint planar subgraphs whose union is GfH.

A very similar argument can be used to provide the same result for the **arboricity** arb(G) of a graph G (the minimum number of line-disjoint acyclic subgraphs whose union equals G).

Proposition 8. Let GfH be a partial function graph, then

(i)   arb(G) < arb(H)   implies   arb(GfH) = arb(H);   and
(ii) arb(G) ≥ arb(H)   implies   arb(GfH) ≤ arb(G) + 1.

Propositions 7 and 8 are of some interest especially since there do not exist efficient algorithms for deciding the value of $t(G)$ or arb(G) for an arbitrary graph G. For example, it is not known (cf. Beineke and Harary [1]) whether the thickness of K_{16} is 3 or 4.

One method of obtaining such algorithms would involve developing reduction procedures, whereby one could reduce a given graph G to a smaller graph G' such that $t(G) = t(G')$ or arb(G) = arb(G'). Proposition 7 and 8 provide hints about the possibilities of such reductions.

4. Functionality. It seems natural to ask if a graph G is not functional, then how close, in some sense, is G to being functional. We next propose one measure of closeness.

As we indicated earlier, any graph can be expressed in many ways as a union of two point disjoint subgraphs together with the set of lines joining the two subgraphs. Furthermore, this set of connecting lines can be expressed as a union of a number of partially defined functions from the points of one subgraph to the points of the other. Let a graph G be so decomposed, we write

$$G = G_1 \cup G_2 \cup \{f_j\} ,$$

where each f_j is a partially defined function from $V(G_1)$ to $V(G_2)$.

The functionality $F(G)$ of a graph G is the minimum number of functions in a set $\{f_j\}$ corresponding to a decomposition $G = G_1 \cup G_2 \cup \{f_j\}$ such that the domain of every function in the set $\{f_j\}$ is all of $V(G_1)$. Clearly, for every graph G, $F(G) \leq$ min deg G.

We now establish a result which bears on the relationship between $F(G)$ and min deg G. By the domain $D\{f_j\}$ of a set of functions $\{f_j\}$ we mean the set of points contained in the domain of at least one function in the set $\{f_j\}$.

Proposition 9. Among all decompositions $G = G_1 \cup G_2 \cup \{f_j\}$ of a graph G,

$$\min | D\{f_j\} | + | \{f_j\} | = \min \deg G + 1.$$

Proof. Let $G = G_1 \cup G_2 \cup \{f_j\}$ be a decomposition for which $|D\{f_j\}| + |\{f_j\}|$ is minimum. Let $|D\{f_j\}| = m$, $|\{f_j\}| = n$ and suppose $m+n <$ min deg G+1. Since $|\{f_j\}| = n$, at least one point u of $V(G_1)$ must be adjacent to exactly n points of $V(G_2)$. Suppose u is also adjacent to less than $m - 1$ points of $V(G_1)$, i.e., deg u ≤ m - 2 + n. Consequently there is a decomposition of $G = \{u\} \cup G - \{u\} \cup \{f_u\}$ for which

$$|D\{f_u\}| + |\{f_u\}| \leq m - 1 + n;$$ a contradiction.

Therefore u must be adjacent to all $m-1$ other points of $V(G_1)$, i.e., $\deg u = m - 1 + n$. But in turn $G = \{u\} \cup G - \{u\} \cup \{f_u\}$ still gives a decomposition for which $|D\{f_u\}| + |\{f_u\}| = m + n = 1 + \deg u$. However by assumption, $m + n < \min \deg G + 1$; i.e., $\deg u < \min \deg G$; another contradiction. Thus $\min |D\{f_j\}| + |\{f_j\}| = \min \deg G + 1$.

5. **Critical Graphs.** Gallai in his two part manuscript [5] presented not only a comprehensive survey of the literature on critical graphs, but a thorough description of the known methods for constructing various classes of critical graphs. In this direction we next present a new method which arises from our study of function graphs. First, consider the 4-critical graph G_4 of Figure 5 which we will express as $G_4 = C_5 \; f \; \overline{W_7}$.

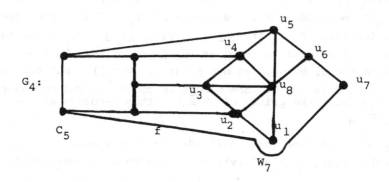

Figure 5

Notice that G_4 essentially consists of the 3-critical five-cycle C_5, the 4-critical wheel of circumference 7 minus one line, $\overline{W_7}$, together with a function f from C_5 to $\overline{W_7}$. To see that G_4 is 4-critical, observe first that G_4 is 4-chromatic, i.e., $\overline{W_7}$ can be 3-colored, but only in such a way that points $u_1, u_3, u_5,$ and u_7 all receive the same color, say c_1. The five-cycle C_5 can of course be 3-colored, but because of the function f connecting C_5 to $u_1, u_3, u_5,$ and u_7, no point of C_5 can be colored c_1 and thus 4-colors are needed to color G_4. To see that G_4 is, in fact, 4-critical one need only consider three cases in regard to removing a line (u,v) from G_4:

(i) $(u,v) \in f$, in this case point $u \in V(C_5)$ can be colored the same as points u_1, u_3, u_5 and u_7 in $G_4 - (u,v)$, the remaining points of C_5 can be colored using only two colors, say c_2 and c_3;

(ii) $(u,v) \in E(C_5)$; but then $C_5 - (u,v)$ can be 2-colored and hence $G_4 - (u,v)$ can be 3-colored;

(iii) $(u,v) \in E(W_7^-)$; this is the most difficult case to handle in theory, but for this particular graph G_4 it can be observed by inspection that for all lines $(u,v) \in E(W_7^-)$, $\chi(G_4 - (u,v)) = 3$.

In an effort to circumvent some of the difficulties which arise in treating case (iii) above, we now present a method for constructing critical graphs which slightly modifies that used to construct G_4 in Figure 5.

<u>Theorem 10.</u> Let G be an $(n-1)$-critical graph, let H be an $(n-2)$-critical graph, and let f be an $(n-2)$ to 1 function from $V(G)$ <u>onto</u> $V(\overline{K}_m)$, $m \geq 2$, i.e., for every point $u \in V(\overline{K}_m)$, $|f^{-1}(u)| \leq n-2$; then the function graph $Gf(\overline{K}_m+H)$ is n-critical.

<u>Proof.</u> $Gf(\overline{K}_m+H)$ is not $(n-1)$-chromatic since every $(n-1)$-coloring of $\overline{K}_m+H$ must color all the points of $\overline{K}_m$ alike. Consequently, since the function f is from <u>all</u> of $V(G)$ <u>onto</u> $V(\overline{K}_m)$, no point of G could be colored the same as any point of $V(\overline{K}_m)$ (every point of G is adjacent, by f, to one point of $V(\overline{K}_m)$). But this means that none of the $(n-1)$-colors needed to color G can be the same as that used to color the points of $V(\overline{K}_m)$; thus n-colors are needed to color $Gf(\overline{K}_m+H)$.

It remains to show that $Gf(\overline{K}_m+H)$ is n-critical, i.e., for every line (u,v), $\chi(Gf(\overline{K}_m+H) - (u,v)) = n-1$. We must consider four cases:

(i) $(u,v) \in E(G)$; since G is, by assumption, $(n-1)$-critical, $\chi(G - (u,v)) = n-2$; thus colors $c_1, c_2, \ldots, c_{n-2}$ can be used to color both G and H; color c_{n-1} can be used to color $\overline{K}_m$;

(ii) $(u,v) \in f$; in this case color H with colors $c_1, c_2, \ldots, c_{n-2}$, color $\overline{K}_m$, including point v, with color c_{n-1}, color point $u \in V(G)$ with c_{n-1}, and color the remaining points of G with colors $c_1, c_2, \ldots, c_{n-2}$ (this is possible since $\chi(G - \{u\}) = n-2$);

(iii) $(u,v) \in E(H)$; since H is, by assumption, $(n-2)$-critical, $\chi(H - (u,v)) = n-3$, therefore color $H - (u,v)$ with colors $c_1, c_2, \ldots, c_{n-3}$, color every point but one of $\overline{K}_m$ with c_{n-1}, color the remaining point of $\overline{K}_m$, say w, with c_{n-2}, color one point, say v, of G adjacent to w with c_{n-1} and color the remaining points of H with colors $c_1, c_2, \ldots, c_{n-2}$. This can always be done since $\chi(G - \{v\}) = n-2$, and at most $n-3$ other points of G besides v, are adjacent to the point w, of $\overline{K}_m$, which is colored c_{n-2}.

(iv)   (u,v) ∈ +, i.e., (u,v) is a line between $\overline{K}_m$ and H;
let u ∈ V$(\overline{K}_m)$, v ∈ V(H); color the points of H - {v} with colors
c_1, c_2, ..., c_{n-3}, color v with c_{n-2}, color u also with c_{n-2},
color the remaining points of V$(\overline{K}_m)$ with c_{n-1}, color one of the
≤ n - 2 points of G adjacent to u with c_{n-1} and color the remaining points of H with colors c_1, c_2, ..., c_{n-2}, as in (iii) a-
bove.

The 4-critical function graphs in Figure 6 are all constructed
according to the method of Theorem 10.

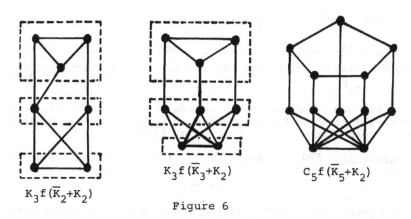

$K_3 f(\overline{K}_3+K_2)$ $C_5 f(\overline{K}_5+K_2)$

$K_3 f(\overline{K}_2+K_2)$

Figure 6

As a further illustration of this method, in addition to
$C_5 f(\overline{K}_5+K_2)$ of Figure 6 one can construct two 4-critical graphs of
the form $C_5 f(\overline{K}_4+K_2)$ and three 4-critical graphs of the form
$C_5 f(\overline{K}_3+K_2)$.

We note, in passing, that for some (n - 1)-critical graphs, the
restriction in Theorem 10 that the function f be at most (n-2) to
1 can be lifted; it is strongly suggested therefore that other improvements of this method for constructing critical function graphs
may be found with more work.

6. _The chromatic numbers of π-graphs._ We will first consider permutation graphs; let us carefully redefine them. A _permutation graph_
G'αG" _of a graph_ G is a graph which consists of two isomorphic
copies G',G" of G together with a set of lines between the copies
which defines a permutation α of V(G), i.e., point u' in G'
is adjacent to point v" in G" if and only if α(u') = v". From
this definition of a permutation graph one can easily establish the
following two results.

Proposition 11. For any graph G and any permutation α, $α^{-1}$ of
V(G), G'αG" ≅ G'$α^{-1}$G".

Proposition 12. Let α,β be any two permutations of V(K_n), then

$K_n'\alpha K_n'' \cong K_n'\beta K_n''$.

Thus from Proposition 12 we see that there is essentially only one permutation graph of K_n; following Chartrand and Harary [4] we denote this by $P(K_n)$. It also follows trivially that there is only one permutation graph on $\overline{K}_n$. Proposition 12 can be extended as follows.

Let $G = (V,E)$ be any graph and let α_1, α_2, ..., α_n be n disjoint permutations of $V(G)$, i.e., for any $u \in V(G)$, $\alpha_i(u) = \alpha_j(u)$ implies $\alpha_i = \alpha_j$. By the permutations graph $G'\alpha_1\alpha_2...\alpha_n G''$ we mean the graph $\bigcup\limits_{i=1}^{n} G'\alpha_i G''$.

Proposition 13. Let $\alpha_1, \alpha_2, ..., \alpha_n$ and $\beta_1, \beta_2, ..., \beta_n$ be two sets of n disjoint permutations of $V(K_m)$, $n < m$; then $K_m'\alpha_1\alpha_2...\alpha_n K_m'' \cong K_m'\beta_1\beta_2...\beta_n K_m''$.

We are now ready to obtain several results concerning chromatic numbers, the first and most obvious of which is an immediate corollary of Brooks' Theorem (cf. Ore [8, p. 226]).

Proposition 14. For any graph G having $p \geq 2$ points and any permutation α of $V(G)$,

$$\chi(G'\alpha G'') \leq p.$$

Proof. Observe only that $\max \deg G \leq p$ and apply Brooks' Theorem. An alternative and more illustrative proof goes as follows. Color G' with $\chi(G) = n$ colors. Let u_1', u_2', ..., u_n' be n points of G' colored with n different colors $c_1, c_2, ..., c_n$, respectively. We proceed to color the second copy G'' in accordance with the permutation α as follows. Color $\alpha(u_i') = \alpha'(u_n')$ with c_1, and for every $i = 1,2,...,n-1$ color $\alpha(u_i') = \alpha(u_i')$ with c_{i+1}. If we now color each of the remaining $p - \chi(G)$ points of G'' with a new color, we will produce a coloring of $G'\alpha G''$ with $\chi(G)+p-\chi(G) = p$ colors.

Corollary 14a. Let $\alpha_1\alpha_2...\alpha_{n-1}$ be disjoint permutations of $V(K_n)$; then

$$\chi(K_n'\alpha_1\alpha_2...\alpha_{n-1}K_n'') = n.$$

Proof. For any $k' \in V(K_n')$ there will exist exactly one $k'' \in V(K_n'')$ such that k' and k'' are not adjacent. Color k' and k'' alike.

The proof of Corollary 14a is equivalent to the assertion that if one removes a 1-factor from K_{2n} the chromatic number of the resulting graph drops to n. This assertion in turn leads us to a result in a different area of graph theory. The bipartity, bip(G), of a graph G is the smallest number of line disjoint bipartite subgraphs whose union equals G.

<u>Theorem 15.</u> If $\chi(G) = 2^n$, then $bip(G) \leq n$.

<u>Proof.</u> Let φ be a homomorphism of G onto K_{2^n}. Clearly K_{2^n} has a one-factor, call it E_{2^n}. Consider $\varphi^{-1}(E_{2^n}) = \{(u,v) \mid (u,v) \in E(G), (u\varphi,v\varphi) \in E_{2^n}\}$. Clearly, the subgraph $B_1 = \langle \varphi^{-1}(E_{2^n}) \rangle$, induced by the set of lines which map into E_{2^n}, is bipartite. Let $G_0 = G$, let $G_1 = G_0 - B_1$, and consider $K_{2^n} - E_{2^n}$. This graph $K_{2^n} - E_{2^n}$ has a homomorphism onto K_{2^n-1}; call it φ_1. Thus, G_1 has a homomorphism $\varphi\varphi_1$ onto K_{2^n-1}.

We now repeat the above process of extracting a 1-factor from K_{2^n-1}, call it E_{2^n-1}, by considering the set of lines of G_1, $(\varphi\varphi_1)^{-1}(E_{2^n-1})$, $B_2 = \langle (\varphi\varphi_1)^{-1}(E_{2^n-1}) \rangle$, etc.

Continuing in this way we obtain successively from G line-disjoint bipartite subgraphs,
$$B_1 = \langle \varphi^{-1}(E_{2^n}) \rangle, \quad B_2 = \langle (\varphi\varphi_1)^{-1}(E_{2^n-1}) \rangle,$$
$$B_3 = \langle (\varphi\varphi_1\varphi_2)^{-1}(E_{2^n-2}) \rangle, \quad \ldots,$$

$B_{n-1} = \langle (\varphi\varphi_1\varphi_2\cdots\varphi_{n-2})^{-1}(E_{2^2}) \rangle$, until finally we see that $G - B_1 - B_2 - \ldots - B_{n-1}$ is itself bipartite, and hence $bip(G) \leq n$.

<u>Corollary 15a.</u> Let $\chi(G) = 2n$; then there exists a bipartite subgraph H of G such that $\chi(G - E(H)) \leq n$.

The following result is due to Chartrand and Frechen [2].

<u>Theorem 16.</u> For any graph G and any permutation α of $V(G)$, $\chi(G'\alpha G'') \leq \left\{\frac{4}{3}\chi(G)\right\}$.

We have seen in Corollary 14a one class of permutation graphs for which $\chi(G'\alpha G'') = \chi(G)$. The next result provides us with a second class.

<u>Theorem 17.</u> For any graph G, $G \neq \overline{K}_n$, and any automorphism α of G, $\chi(G'\alpha G'') = \chi(G)$.

<u>Proof.</u> Let $\chi(G) = n$ and let ϕ' be a homomorphism of G' onto K_n', where $V(K_n') = \{k_1', k_2', \ldots, k_n'\}$. In a natural way ϕ' can be used to define another homomorphism ϕ'' of G'' onto a second complete graph K_n'', where $V(K_n'') = \{k_1'', k_2'', \ldots, k_n''\}$, as follows:

If $\phi'(\alpha^{-1}(v")) = k_i'$, then define $\phi"(v") = k_i"$. [We view the automorphism α here as a mapping from $v' \leftrightarrow v"$.] The homomorphism $\phi"$ is equivalent to (although, precisely, not <u>equal</u> to) the composition of the automorphism α^{-1} with the homomorphism ϕ', cf. Figure 7.

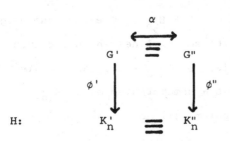

Figure 7.

Now we define a graph H as follows:
$V(H) = \{u_1, u_2, \ldots, u_n, v_1, v_2, \ldots, v_n\}$, and

$E(H) = \{(u_i,u_j) \mid i,j = 1,2,\ldots,n; \ i \neq j\} \cup \{(v_i,v_j) \mid i,j = 1,2,\ldots,n; \ i \neq j\} \cup \{(u_i,v_j) \mid \exists u \in V(G'), \ \exists v \in V(G"), \ \phi'(u) = k_i, \ \phi"(v) = k_j \text{ and } \alpha(u) = v\}.$

The graph H is, in fact, a permutation graph of the form $P(K_n)$, which is a homomorphic image of $G'\alpha G"$ under the combined homomorphisms ϕ' and $\phi"$. Thus we define a homomorphism $\phi' \cup \phi"$ of $G'\alpha G"$ onto H as follows:

for $u \in V(G')$ if $\phi'(u) = k_i'$, let $(\phi'\cup\phi")(u) = u_i$.

for $v \in V(G")$ if $\phi"(v) = k_j"$, let $(\phi'\cup\phi")(v) = v_j$.

Since $\phi'\cup\phi"$ defines a homomorphism of $G'\alpha G"$ onto H, we have $\chi(G'\alpha G") \leq \chi(H)$.

Observe next that for a given $u_i \in V(H)$ there exists exactly one point v_j for which $(u_i,v_j) \in E(H)$, and conversely, for a given $v_j \in V(H)$ there exists exactly one point u_i for which $(u_i,v_j) \in E(H)$. For suppose (u_i,v_j) and $(u_i,v_r) \in E(H)$. Then by the definition of $E(H)$ there exist $u_1,u_2 \in V(G')$, $v_1,v_2 \in V(G")$, $\phi'(u_1) = \phi'(u_2) = k_i'$, $\phi"(v_1) = k_j"$, $\phi"(v_2) = k_r"$ and $\alpha(u_1) = v_1$, $\alpha(u_2) = v_2$. But by definition of $\phi"$, which maps $G"$ onto $K_n"$, $\phi"(v_1) = \phi"(v_2)$, and consequently $(\phi'\cup\phi")(v_1) = (\phi'\cup\phi")(v_2)$, but $(\phi'\cup\phi")(v_1) = v_j$ and $(\phi'\cup\phi")(v_2) = v_r$, thus $v_j = v_r$.

Hence $H \cong P(K_n)$. But by Corollary 13a, $\chi(P(K_n)) = n$. Thus

$\chi(G'\alpha G")\leq\chi(H)=n$. But trivially since $\chi(G)=n$, $\chi(G'\alpha G")\geq n$; therefore $\chi(G'\alpha G")=n=\chi(G)$, completing the proof.

A slight generalization of the proof of Theorem 17 yields the following:

<u>Corollary 17a</u>. If ϕ is a homomorphism from a graph G_1 to G_2, $G_1\neq\overline{K}_n$, then for the homomorphism graph $G_1\phi G_2$ we have

$$\chi(G_1\phi G_2)=\chi(G_2) .$$

The next result was suggested by Propositions 11 and 12 and Theorem 16.

<u>Theorem 18</u>. Let G,H be two graphs and let α,β be two (partial) functions from $V(G)$ to $V(H)$ for which there exist automorphisms $g\in\Gamma(G)$, $h\in\Gamma(H)$ such that $h\alpha=\beta g$; then $G\alpha H\cong G\beta H$.

<u>Proof</u>. We define the isomorphism i from $G\alpha H$ to $G\beta H$ as follows: for $u_G\in V(G\alpha H)$, $i(u_G)=g(u_G)\in V(G\beta H)$, and for $u_H\in V(G\alpha H)$, $i(u_H)=h(u_H)\in V(G\beta H)$. Thus i essentially maps G in $G\alpha H$ to G in $G\beta H$ according to the automorphism g of G; and i maps H in $G\alpha H$ to H in $G\beta H$ according to the automorphism h of H. Since the mapping i is clearly 1-1 and onto, it only remains to show that it is a homomorphism, i.e., that

(i) $u_G\ \text{adj}_\alpha\ v_G\Rightarrow i(u_G)\ \text{adj}_\beta\ i(v_G)$;

(ii) $u_H\ \text{adj}_\alpha\ v_H\Rightarrow i(u_H)\ \text{adj}_\beta\ i(v_H)$;   and

(iii) $u_G\ \text{adj}_\alpha\ \alpha(u_G)\Rightarrow i(u_G)\ \text{adj}_\beta\ i(\alpha(u_G))$.

Now (i) and (ii) follow because of the automorphisms g and h, respectively, and (iii) follows since $i(u_G)=g(u_G)$, $i(\alpha(u_G))=h(\alpha(u_G))$, for by assumption, $h\alpha(u_G)=\beta g(u_G)$, and by definition of $G\beta H$, $g(u_G)\ \text{adj}_\beta(g(u_G))$.

The condition in Theorem 18 that $h\alpha=\beta g$ is equivalent to saying that the diagram in Figure 8 commutes.

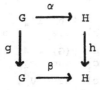

Figure 8

<u>Corollary 18a</u>. Let $G'\alpha G"$ and $G'\beta G"$ be two permutation graphs for which there exist automorphisms $g,h\in\Gamma(G)$ such that $g\alpha=\beta h$; then $G'\alpha G"\cong G'\beta G"$.

<u>Corollary 18b</u>. Let $G'\alpha G"$ and $G'\beta G"$ be two permutation graphs for which there exists an automorphism $g\in\Gamma(G)$, such that

$g \alpha g^{-1} = \beta$; then $G' \alpha G'' \cong G' \beta G''$.

Corollary 18c. For any two automorphisms $g, h \in \Gamma(G)$, $G'gG'' \cong G'hG''$.

Corollary 18d. For any two automorphisms $g, h \in \Gamma(G)$ and any permutation α of $V(G)$, $G' \alpha G'' \cong G'g\alpha hG''$.

In their paper on planar permutation graphs, Chartrand and Harary [4] observed that sometimes many different permutations α can produce the same permutation graph $G' \alpha G''$ (up to isomorphism). They never addressed themselves however to the fascinating question: for a given graph G how many distinct permutation graphs $G' \alpha G''$ are there?

We see from Corollary 18d that each of the <u>double cosets</u> (cf. Hall [6, p. 14]) $\Gamma(G) \alpha \Gamma(G)$ of the symmetric group S_n, $\alpha \in S_n$, determines one permutation graph of G. And we see from Corollary 18c that one of these double cosets corresponds exactly to the automorphism group of G, $\Gamma(G)$. We also observe from Proposition 11 that two distinct double cosets, namely $\Gamma(G) \alpha \Gamma(G)$ and $\Gamma(G) \alpha^{-1} \Gamma(G)$, can produce the same permutation graph. Thus it would seem that the number of distinct permutation graphs $G' \alpha G''$ could be considerably less that the number of double cosets $\Gamma(G) \alpha \Gamma(G)$.

The following result represents still another refinement of the bound in Theorem 16.

Proposition 19. If α is a transposition of $V(G)$ and $\chi(G) \leq 3$ then $\chi(G' \alpha G'') \leq \chi(G) + 1$, and if $\chi(G) \geq 4$ then $\chi(G' \alpha G'') = \chi(G)$.

Proof. The bound for $\chi(G) \leq 3$ follows from Theorem 16; permutation graphs which meet these bounds are shown in Figure 9. Let $\chi(G) > 4$ and let α be a transposition; i.e., $\alpha(u) = v$, $\alpha(v) = u$, and for all other points $w \in V(G)$, $\alpha(w) = w$. Color G' with $\chi(G) = n$ colors and let $V_1', V_2', \ldots, V_n'$, colored $c_1, c_2, \ldots, c_n$ respectively, be the n color classes. Let $V_1'', V_2'', \ldots, V_n''$ be the corresponding color classes of G''. We have to consider only two cases: (i) $u, v \in V_i'$ for some i; in this case color the classes $V_1'', V_2'', \ldots, V_n''$ with colors $c_2, c_3, \ldots, c_n, c_1$, respectively; (ii) $u \in V_i'$, $v \in V_j'$; in this case color classes V_i'' and V_j'' with two different colors. We can do this since $\chi(G) \geq 4$ and there will exist two extra colors not the same as c_i and c_j. It only remains to color the remaining $n - 2$ color classes V_m'', but this can always be done in such a way that no class V_m'' is colored the same as its corresponding class V_m'.

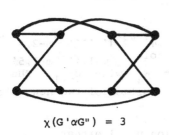
$\chi(G'\alpha G'') = 3$

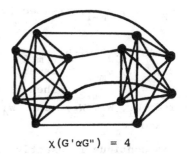
$\chi(G'\alpha G'') = 4$

Figure 9

The method used to establish Proposition 19 can be extended to produce the following:

<u>Corollary 19a</u>. If α is a permutation of $V(G)$ which can be expressed as a product of two transpositions, then for $\chi(G) \leq 3$, $\chi(G'\alpha G'') \leq \chi(G)+1$, and for $\chi(G) \geq 4$, $\chi(G'\alpha G'') = \chi(G)$.

A brief attempt was made to establish the above result when α can be expressed as a product of three transpositions, but the number of distinct cases to be considered grew so large that the effort was terminated.

We now close our discussion by providing an upper bound for the chromatic number of function graphs, a bound which bears a close relationship to that given in Theorem 16 for permutation graphs.

<u>Theorem 20</u>. Let $\chi(G) = m$, $\chi(H) = n$, then for any function graph GfH,

(i) if $2m \leq n$, then $\chi(GfH) = \chi(H) = n$

(ii) if $2m > n$, then $\chi(GfH) \leq m + \left\{\frac{n}{2}\right\}$.

<u>Proof</u>. Color G with m colors and let $V_1, V_2, \ldots, V_m$ be the corresponding color classes. Color H with n colors and let $W_1, W_2, \ldots, W_n$ be the corresponding color classes. Define next a new partition $\{V_{ij}\}$ of $V(G)$, where V_{ij} is the set of those points of V_i adjacent, according to f, to a point of W_j. Now, from the given colorings of G and H we proceed to a coloring of GfH by recoloring the subsets V_{ij} and W_k.

To begin with, color $V_{11}, V_{21}, \ldots, V_{m,1}$ with colors $c_1, c_2, \ldots, c_m$, respectively; color W_1 with c_{m+1}, also color subsets $V_{m,2}, V_{m,3}, \ldots, V_{m,n}$ with c_{m+1}.

Next, color $V_{1,2}, V_{2,2}, \ldots, V_{m-1,2}$ with $c_1, c_2, \ldots, c_{m-1}$, respectively; color W_2 with c_m. Thus, so far, we have colored

W_1 and W_2 with <u>one</u> additional color beyond $c_1, c_2, \ldots, c_m$.

Continuing in this manner, color $V_{1,3}, V_{2,3}, \ldots, V_{m-1,3}$ with $c_1, c_2, \ldots, c_{m-1}$, respectively; $V_{m,3}$ has already been colored c_{m+1}.

Color W_3 with c_{m+2}. Color $V_{1,4}, V_{2,4}, \ldots, V_{m-2,4}$ with $c_1, c_2, \ldots, c_{m-2}$, respectively, and color $V_{m-1,4}, V_{m-1,5}, \ldots, V_{m-1,n}$ with c_{m+2}. Then we can color W_4 with c_{m-1}. Thus we have colored W_1, W_2, W_3, and W_4 by the addition of two new colors c_{m+1} and c_{m+2}.

We continue in exactly this fashion until either
(i) $2m > n$, and we exhaust all the sets W_j, at which point we will have used $m + \left\{\frac{n}{2}\right\}$ colors, i.e., one new color for every other set W_j, i.e., W_1, W_3, W_5, etc., or

(ii) $2m \leq n$, at which point we shall have added m new colors for $W_1, W_2, \ldots, W_{2m}$, and from then on we can add one new color for each of the remaining sets $W_{2m+1}, W_{2m+2}, \ldots, W_n$; in this way we shall use $m + m + n-2m = n$ colors.

The diagrams in Figure 10 illustrate the coloring procudure used in proving Theorem 20.

(a)			(b)			(c)		
$\chi(G) = 2$ $\chi(H) = 6$			$\chi(G) = 4$ $\chi(H) = 5$			$\chi(G) = 4$ $\chi(H) = 3$		
V_1 V_2			V_1 V_2 V_3 V_4			V_1 V_2 V_3 V_4		
1 2	3	W_1	1 2 3 4	5	W_1	1 2 3 4	5	W_1
1 3	2	W_2	1 2 3 5	4	W_2	1 2 3 5	4	W_2
1 3	4	W_3	1 2 3 5	6	W_3	1 2 3 5	6	W_3
4 3	1	W_4	1 2 6 5	3	W_4			
4 3	5	W_5	1 2 6 5	7	W_5			
4 3	6	W_6						

Figure 10

<u>Corollary 20a.</u> For any graph G and any function f from $V(G)$ to $V(G)$,

$$\chi(GfG) \leq \left\{\frac{3}{2} \chi(G)\right\}.$$

The author gratefully acknowledges discussions with Dennis Geller and Arthur Fleck during the early development of this material.

REFERENCES

1. L.W. Beineke and F. Harary. The thickness of the complete graph, Canad. J. Math. 17 (1965), 850-859.

2. G. Chartrand and J.B. Frechen, On the chromatic number of permutation graphs, Proof Techniques in Graph Theory, (F. Harary, ed.), Academic Press, New York, 1969 (to appear).

3. G. Chartrand, D. Geller, S. Hedetniemi, Graphs with forbidden subgraphs, J. Combinatorial Theory, (to appear).

4. G. Chartrand and F. Harary, Planar permutation graphs, Ann. Inst. Henri Poincaré, Vol. IIIB (1967), 433-438.

5. T. Gallai, Kritische Graphen I, II, Magyar Tud. Akad. Mat. Kutató Int. Közl. 8 (1963), 165-192, and (1964) 373-395.

6. M. Hall, The theory of groups, Macmillan, New York, 1959, p. 14.

7. F. Harary and G. Prins, The block-cutpoint-tree of a graph, Publ. Math. Debrecen 13 (1966), 103-107.

8. O. Ore, Theory of graphs, Amer. Math. Soc. Colloq. Publ. 38, Providence, 1962.

RANK 3 GRAPHS

Marshall D. Hestenes, Michigan State University

If G is a transitive permutation group on a set Ω and G_a is the subgroup of G stabilizing the point $a \in \Omega$, then the <u>rank</u> of G is the number of orbits of G_a in Ω. Thus rank 2 means double transitivity. The fact that many of the known finite simple groups have representations as rank 3 groups, even though some of them do not have doubly transitive representations, recommends this class as a possible source of new groups [1]. The recent discovery of new simple groups as rank 3 groups by D.G. Higman and C.C. Sims [2], M. Suzuki [5], and J.E. McLaughlin [4] bears this out. Consequently, D.G. Higman and the author have embarked on a systematic study of non-solvable primitive rank 3 groups of even order. The following is a brief description of the study undertaken with particular emphasis on the role that graph theory has played here.

Suppose G is a rank 3 group on Ω, and for $a \in \Omega$, G_a has orbits $\{a\}$, $\Delta(a)$, and $\Gamma(a)$. The notation is chosen so that $\Delta(a^g) = \Delta(a)^g$ and $\Gamma(a^g) = \Gamma(a)^g$. Denote the lengths of Ω, $\Delta(a)$ and $\Gamma(a)$ by n, k and ℓ, respectively. If $|G|$ is even, then $b \in \Delta(a)$ if and only if $a \in \Delta(b)$, and we say that Δ is self-paired; while if $|G|$ is odd, then $b \in \Delta(a)$ implies that $a \in \Gamma(b)$.

Let G_Δ be a graph with Ω as its vertex set and with a vertex b adjacent to a if $b \in \Delta(a)$. Since G_Δ is a regular graph or digraph, and G_Δ is a graph if and only if Δ is self-paired, we assume that $|G|$ is even. The graph G_Γ defined similarly turns out to be the complement of G_Δ.

A graph admitting a (primitive) rank 3 automorphism group is, by definition, a (<u>primitive</u>) <u>rank 3 graph</u>. Of course, G_Δ and G_Γ are rank 3 graphs. Moreover, since a rank 3 graph is primitive if and only if both the graph and its complement are connected, we assume that both G_Δ and G_Γ are connected. Let

$$|\Delta(x) \cap \Delta(y)| = \begin{cases} \lambda & \text{if } y \in \Delta(x) \\ \mu & \text{if } y \in \Gamma(x) . \end{cases}$$

Then G_Δ is a graph such that

(a) G_Δ is regular of degree k

(b) each edge in G_Δ belongs to exactly λ triangles, and

(c) every nonadjacent pair of vertices are joined by exactly μ paths of length 2.

Thus G_Δ is a strongly regular graph. However, not all strongly regular graphs are rank 3 graphs.

Call the set $\{n,k,\ell,\lambda,\mu\}$ the parameters of G_Δ. Using only the strongly regular properties of G_Δ, we obtain conditions on the parameters, and then program the computer to find the possible sets of parameters satisfying these conditions for $n < 1000$. Our goal is to examine the list for existence and uniqueness of groups for each set of parameters. Techniques are still being developed to handle these problems.

Graph theory can be useful in this connection. For example, the methods of Hoffman [3] and others may be used to completely determine the strongly regular graphs whose adjacency matrix has minimum eigenvalue -2. It is then easy to determine which of these admit rank 3 automorphism groups, and then identify the groups.

If u and v are distinct vertices of a graph G, then two subgraphs of G are said to be of the same type with respect to (x,y) if they both contain x and y, and there is an isomorphism of one onto the other, mapping x onto x and y onto y. For every $t \geq 2$, the rank 3 graphs satisfy the t-vertex condition: Given distinct vertices x and y, the number of t-vertex subgraphs of a given type with respect to (x,y) depends only on whether xy is an edge of the graph or not. We have already used the t-vertex condition for $t \leq 3$. Assuming the condition for $t = 4$, we are able to introduce new parameters and determine conditions on them. Again, this has been programmed for the computer.

Finally, let us assume that a known rank 3 group has a rank 3 extension G. Then the points at distance one or two from a given point in G_Δ again form a rank 3 graph. This assumption imposes certain conditions on the parameters, and the computer is used to find out when this is possible. Then we attempt to find the group, if it exists. This method is very promising since it is basically that used to find the three new simple groups mentioned earlier.

At this time there are only a few parameter sets with $n < 100$ where we have no information. Moreover, we have acquired a fairly extensive list of rank 3 representations of known groups. Of course this is a rich source of examples of strongly regular graphs.

REFERENCES

1. D.G. Higman, Finite permutation groups of rank 3, <u>Math. Z.</u>, 86 (1964), 145-156.

2. D.G. Higman and C.C. Sims, A simple group of order 44,352,000, <u>Math. Z.</u>, 105 (1968), 110-113.

3. A.J. Hoffman, On the uniqueness of the triangular association scheme, <u>Ann. Math. Stat.</u>, 31 (1960), 492-497.

4. J.E. McLaughlin, A simple group of order $275|U_4(3)|$, to appear.

5. M. Suzuki, A simple group of order 448,345,497,600, to appear.

VARIATIONS ON A THEOREM OF PÓSA

Hudson V. Kronk, SUNY at Binghamton

The problem of characterizing hamiltonian graphs is unsolved and seems to be very difficult. There are, however, several sufficient conditions for a graph to be hamiltonian. In [10] Pósa proved the following result.

<u>Theorem 1</u>. Let G be a graph with $p \geq 3$ points satisfying the following conditions: (1) for every integer k such that $1 \leq k < \frac{(p-1)}{2}$, the number of points of degree not exceeding k is less than k, (2) the number of points of degree not exceeding $\frac{(p-1)}{2}$ does not exceed $\frac{(p-1)}{2}$. Then G is hamiltonian.

This theorem has several interesting corollaries which we now state.

<u>Corollary 1a</u>. Let G be a graph with $p \geq 3$ points such that for every integer k with $1 \leq k < \frac{p}{2}$, the number of points of degree not exceeding k is less than k. Then G is hamiltonian.

Our next corollary was first discovered by Ore [8].

<u>Corollary 1b</u>. A graph G on $p \geq 3$ points is hamiltonian if $\deg u + \deg v \geq p$ for all nonadjacent points u and v in G.

<u>Proof</u>. Let $1 \leq k < \frac{p}{2}$ and let t denote the number of points of G whose degree does not exceed k. These t points form a complete subgraph T; for if any two of these were not adjacent, there would exist two nonadjacent points the sum of whose degrees would be less than p. This implies that $t \leq k + 1$. We cannot have $t = k + 1$, for then each point of T is adjacent only to points of T, and if $u \in T$ and $v \in G - T$, then $\deg u + \deg v \leq k + (p - k - 2) = p - 2$, which is a contradiction. We cannot have $t = k$ either, for in this case each point of T is adjacent to at most one point not in T. But since $t = k < \frac{p}{2}$, there exists a point $w \in G - T$ adjacent to no point of T. Then if $u \in T$, $\deg u + \deg w \leq k + (p - k - 1) = p - 1$, which is again a contradiction. Hence $t < k$, so that by Corollary 1a, G is hamiltonian.

If G has no points of degree less than $\frac{p}{2}$, then G satisfies the hypothesis of Ore's theorem. This provides another sufficient condition due to Dirac [6].

<u>Corollary 1c</u>. Let G be a graph with $p \geq 3$ points such that $\deg v \geq \frac{p}{2}$ for all points v of G. Then G is hamiltonian.

Finally we state a sufficient condition involving the number of lines of a graph. This result is also due to Ore [9].

Corollary 1d. Let G be a graph having $p \geq 3$ points and
$q \geq \dfrac{(p^2 - 3p + 6)}{2}$ lines. Then G is hamiltonian.

Proof. Let u and v be any two nonadjacent points of G. The
maximum number of lines in G incident with neither u nor v is
$\binom{p-2}{2} = \dfrac{(p^2 - 5p + 6)}{2}$. Therefore $q \leq \dfrac{(p^2 - 5p + 6)}{2}$ + deg u + deg v.
But $q \geq \dfrac{(p^2 - 3p + 6)}{2}$ implies that deg u + deg v $\geq$ p so that by
Corollary 1b, G is hamiltonian.

All of the previously stated results are "sharp." For example,
to see that Pósa's theorem is best possible let
$1 \leq k < \dfrac{p-1}{2}$ and consider the graph G having one cutpoint and two
blocks, one of which is K_{k+1} and the other K_{p-k}. This graph is
not hamiltonian, but it violates the theorem only in that it has exactly k points of degree k. If p is the odd integer 2k + 1,
then the complete bipartite graph $K_{k,k+1}$ is not hamiltonian and
violates the theorem only in that it has exactly $\dfrac{p-1}{2}$ + 1 points of
degree not exceeding $\dfrac{p-1}{2}$.

Our principle goal in this note is to present other "Pósa type"
theorems which were first suggested by a "Dirac type" theorem.

A graph G is said to be n-connected if the removal of fewer
than n points from G neither disconnects it nor reduces it to the
trivial graph. In [1] Chartrand and Harary proved the following result:

Theorem 2. Let G be a graph with p points and let $1 \leq n < p$.
If deg v $\geq \dfrac{(p + n - 2)}{2}$ for all points v of G, then G is n-
connected.

We now state a "Pósa type" improvement, a proof of which can be
found in [2].

Theorem 3. Let G be a graph with p points and let $1 \leq n < p$.
The following conditions are sufficient for G to be n-connected:

> (1) for every integer k such that $n - 1 \leq k < \dfrac{(p + n - 3)}{2}$,
> the number of points of degree not exceeding k does not
> exceed k + 1 - n,

> (2) the number of points of degree not exceeding $\dfrac{(p + n - 3)}{2}$
> does not exceed p - n.

We remark that since every hamiltonian graph is 2-connected,
Pósa's conditions are also sufficient for a graph to be 2-connected,
however, the conditions in Theorem 3 for 2-connectedness are sharper.

In [5], a graph G was defined to <u>randomly</u> <u>hamiltonian</u> if a
spanning path always results upon starting at any point of G and
successively proceeding to any adjacent point not yet encountered and
where the endpoints of this path are adjacent. These graphs were
characterized in [5] as being either cycles, complete graphs or regular complete bipartite graphs. Clearly a graph G is randomly hamiltonian if and only if every path of length k, 0 ≤ k ≤ p - 1, is
contained in a hamiltonian cycle of G. (A point is considered here
as being a path of length zero.) This suggests the following definition.

<u>Definition</u>. A graph G with p ≥ 3 points is <u>k-path hamiltonian</u>,
0 ≤ k ≤ p - 2, if every path of length not exceeding k lies on
some hamiltonian cycle of G.

Analogously, one can define <u>k-line hamiltonian</u> graphs as those
graphs G having the property that any path system of length not exceeding k lies on a hamiltonian cycle of G. (A path system in G
is a subgraph whose components are paths. The length of a path system is the sum of the lengths of its paths.) These two concepts
clearly agree for k = 0,1, and p - 2, but they are not equivalent
in general. For example, the graph G shown in Figure 1 is 2-path
hamiltonian but not 2-line hamiltonian, the lines x and y not being contained in any hamiltonian cycle of G.

In [11], Pósa proved the following "Dirac type" theorem.

<u>Theorem 4</u>. A graph G on p ≥ 3 points is k-line (and therefore
k-path) hamiltonian, 0 ≤ k ≤ p - 2, if $\deg v \geq \frac{(p + k)}{2}$ for all
points v of G.

The "Pósa type" improvement of this theorem of Pósa is the
following:

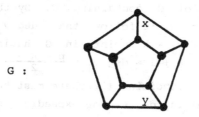

G :

Figure 1: <u>A 2-path hamiltonian graph</u>

<u>which is not 2-line hamiltonian.</u>

<u>Theorem 5</u>. Let G be a graph on p points and let 0 ≤ k ≤ p - 2.
The following conditions are sufficient for G to be k-line (and
therefore k-path) hamiltonian:

(1) for all integers j with $k + 1 \leq j < \frac{(p + k - 1)}{2}$, the
number of points of degree not exceeding j is less than
j - k,

(2) the number of points of degree not exceeding $\frac{(p + k - 1)}{2}$

does not exceed $\frac{(p - k - 1)}{2}$.

__Proof.__ Assume that G satisfies the hypothesis of the theorem but contains a path system W of length not exceeding k which is not contained in a hamiltonian cycle. We may assume, without loss of generality, that G becomes k-line hamiltonian whenever any new line is added to G, for if G did not originally have this property we could add suitable lines until a graph G_1 is obtained with this property such that G_1 also satisfies the hypothesis of the theorem.

Let v_1 and v_p be two nonadjacent points of G such that (1) deg $v_1 \leq$ deg v_p and (2) deg v_1 + deg v_p is as large as possible. If we now add the line $v_1 v_p$ to G we obtain a k-line hamiltonian graph G^*. Let C be a hamiltonian cycle of G^* which contains the path system W. Clearly C must include the line $v_1 v_p$ and hence v_1 and v_p are the endpoints of a spanning path $P = (v_1, v_2, \ldots, v_p)$ in G which also contains the path system W. If v_i, $2 \leq i < p$, is adjacent to v_1 and if $v_{i-1} v_i$ is not in W, then $v_{i-1} v_p$ is not in G. For otherwise, $(v_1, v_i, v_{i+1}, \ldots, v_p, v_{i-1}, v_{i-2}, \ldots, v_1)$ would be a hamiltonian cycle of G containing the path system W. Since at most k lines of P belong to the path system W, it follows that there are at least deg $v_1 - k$ points in G which are nonadjacent to v_p. Therefore,

$$\text{deg } v_1 \leq \text{deg } v_p \leq (p - 1) - (\text{deg } v_1 - k)$$

so that deg $v_1 \leq \frac{(p + k - 1)}{2}$. Furthermore, whenever v_i is adjacent to v_1 and $v_{i-1} v_p$ is not in W, $(v_{i-1}, v_{i-2}, \ldots, v_1, v_i, v_{i+1}, \ldots, v_p)$ is a spanning path of G containing W. By the manner in which v_1 and v_p were chosen, it follows that deg $v_{i-1} \leq$ deg v_1. Thus there are at least deg $v_1 - k$ points in G having degree not exceeding deg v_1. We have $k + 1 \leq$ deg $v_1 \leq \frac{p + k - 1}{2}$. However, if deg $v_1 < \frac{p + k - 1}{2}$, then by condition (1) there must be less than deg $v_1 - k$ points in G having degree not exceeding deg v_1. On the other hand, if deg $v_1 = \frac{(p + k - 1)}{2}$, then deg $v_p = \frac{p + k - 1}{2}$ so that there are at least deg $v_1 - k + 1 = \frac{p - k + 1}{2}$ points in G having degree not exceeding deg $v_1 = \frac{p + k - 1}{2}$. This, however, violates condition (2). Having been led to a contradiction, we conclude that the theorem is true.

Analogous to Corollary 1a we have the following result [7].

__Corollary 5a.__ Let G be a graph on $p \geq 3$ points and let $0 \leq k \leq p - 3$. If for every integer j with $k + 1 \leq j < \frac{(p + k)}{2}$, the number of points of degree not exceeding j is less than j - k, then G is k-line hamiltonian.

In conclusion, we remark that all of the results are, in a sense, best possible. Also, other special classes of "highly" hamiltonian graphs have recently been defined [3,4] and for each of these classes a "Pósa type" theorem analogous to Theorem 5 holds.

REFERENCES

1. G. Chartrand and F. Harary, Theory of Graphs (P. Erdös and G. Katona, eds.) Akadémiai Kiadó, Budapest, 1968, 61-63.

2. G. Chartrand, S.F. Kapoor, and H.V. Kronk, A sufficient condition for n-connectedness of graphs, Mathematika, 15 (1968), 51-52.

3. G. Chartrand, S.F. Kapoor, and H.V. Kronk, A generalization of hamiltonian-connected graphs, to appear.

4. G. Chartrand, S.F. Kapoor, and D.R. Lick, n-hamiltonian graphs, J. Combinatorial Theory, to appear.

5. G. Chartrand and H.V. Kronk, Randomly traceable graphs, J. SIAM Appl. Math., 16 (1968), 696-700.

6. G.A. Dirac, Some theorems on abstract graphs, Proc. London Math. Soc., 2 (1952), 69-81.

7. H.V. Kronk, Generalization of a theorem of Pósa, Proc. Amer. Math. Soc., to appear.

8. O. Ore, Note on hamilton circuits, Amer. Math. Monthly, 65 (1958), 611.

9. O. Ore, Arc coverings of graphs, Ann. Mat. Pura Appl., 55 (1961), 315-322.

10. L. Pósa, A theorem concerning hamilton lines, Magyar Tud. Akad. Mat. Kutató Int. Közl. 7 (1962), 225-226.

11. L. Pósa, On the circuits of finite graphs, Magyar Tud. Akad. Mat. Kutató Int. Közl., 8 (1963), 355-361.

CRITICALLY AND MINIMALLY n-CONNECTED GRAPHS[1]

Don R. Lick, Western Michigan University

1. <u>Introduction</u>. The <u>connectivity</u> $\kappa(G)$ of a graph G is the minimum number of points whose removal either disconnects G or reduces it to the trivial graph consisting of a single point. A graph G is said to be <u>n-connected</u> if $\kappa(G) \geq n$. Analogously, the <u>line-connectivity</u> $\lambda(G)$ of a graph G is the minimum number of lines whose removal disconnects it. A graph G with $\lambda(G) \geq n$ is said to be <u>n-line-connected</u>. If $\mu(G)$ denotes the minimum degree of the graph G, then the inequalitiy $\kappa(G) \leq \lambda(G) \leq \mu(G)$ is well known for connected graphs.

Also, if G is a graph with $\kappa(G) = 1$, then there exists a point v of G (in fact, two points) such that $\kappa(G - v) = 1$. Furthermore, if G is not a tree, then there is a line x of G such that $\kappa(G - x) = 1$. Such statements cannot be made in general, however, for graphs G with $\kappa(G) \geq 2$. Recently, Dirac [4] and Plummer [10] proved that if G is a graph with $\kappa(G) = 2$ and $\mu(G) > 2$, then there exists a line x of G such that $\kappa(G - x) = 2$. More recently, Kaugars [7] showed that such a graph G also contains a point v such that $\kappa(G - v) = 2$. These two results involve only the connectivity and the minimum degree of the graph G. A natural question in this context seems to be the following: Is there some condition that can be placed on the minimum degree of an n-connected graph which assures that it contains a line x and a point v such that the graphs G - x and G - v are n-connected? This paper investigates these questions.

We can rephrase the above question as follows: What n-connected graphs G have $\kappa(G - v) = n - 1$ for each point v of G, and $\kappa(G - x) = n - 1$ for each line x of G? This leads to the following definitions. A graph G is said to be <u>critically n-connected</u> if $\kappa(G) = n$ and $\kappa(G - v) = n - 1$ for each point v of G; and G is said to be <u>minimally n-connected</u> if $\kappa(G) = n$, and for each line x of G, $\kappa(G - x) = n - 1$. Thus if the graphs which are critically n-connected or minimally n-connected are determined, then the above problem will be solved. In [6] Halin showed that no minimally n-connected graph G has $\mu(G) > n$. Here we prove that for every critically n-connected graph G, $\mu(G) < (3n - 1)/2$.

There are analogous concepts where connectivity is replaced by line-connectivity. A graph G is said to be <u>critically n-line-connected</u> if $\lambda(G) = n$ and $\lambda(G - v) = n - 1$ for each point v of G; and G is said to be <u>minimally n-line-connected</u> if $\lambda(G) = n$, and for each line x of G, $\lambda(G - x) = n - 1$. In section 3, we show that there does not exist a minimally n-line-connected graph G with

---

[1]Work supported in part by grants from the National Science Foundation (GP-9435) and Western Michigan University (Faculty Research Fellowship).

$\mu(G) > n$.

From the above definitions, we note that "critical" refers to the removal of a point from a graph, while "minimal" refers to the removal of a line.

2. **Critically n-connected graphs**. The concepts of critically n-con-nectedness and minimally n-connectedness are independent in the sense that neither property implies the other. For example, the graph G_1 of Figure 1 is critically 2-connected but not minimally 2-connected, while the graph $G_2 = K(2,3)$ of Figure 1 is minimally 2-connected and not critically 2-connected.

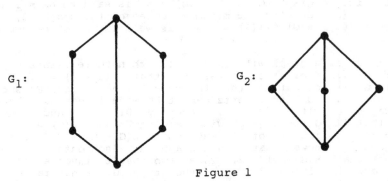

G_1: G_2:

Figure 1

It is easily seen that the graph $K(n,n+1)$ is minimally n-con-nected but not critically n-connected, while for $n \geq 3$ the graph $K(n,n)$ with a line added is critically n-connected but not minimally n-connected.

We conclude our preliminary remarks by pointing out that any n-connected graph always contains a critically n-connected subgraph and a minimally n-connected subgraph.

Theorem 1. There exists no critically n-connected graph with $\mu(G) \geq (3n-1)/2$.

Before proving Theorem 1, we remark that the case $n = 2$ was proved in [7] and the general case was handled in [3] and [8].

Proof. Suppose the theorem to be false so that there exists a graph G of order p having $\kappa(G) = n$ and $\mu(G) \geq (3n-1)/2$ such that for every $v \in V(G)$, $\kappa(G-v) = n - 1$. We note that since $\mu(G) \geq (3n-1)/2$, G is not complete. This implies that every vertex of G belongs to some n-cut set of G.

Among all n-cut sets S' of G, let S be one such that $G - S$ contains a component G_1 of smallest order; denote the order of G_1 by m. Furthermore, let $G_2 = G - S - V(G_1)$.

Let $v \in V(G_1)$ and $u \in V(G_2)$. By a result of Whitney [11] there exist n disjoint u-v paths in G; necessarily, each such path contains precisely one vertex of S. Hence there exist n disjoint paths joining u and S (and also v and S).

Let $w \in V(G_1)$, and let S* be an n-cut set of G containing w. Define $G* = G - S*$ and, furthermore, let $V_1 = V(G_1) \cap S*$, $V_2 = V(G_2) \cap S*$, and $V_3 = S \cap S*$, where $|V_i| = n_i$, i = 1, 2, 3. We note that $n_1 + n_2 + n_3 = n$ and $n_1 \geq 1$.

We now show that $n_2 \geq n_1$. If $S* \supseteq V(G_2)$, then this is obvious. Assume therefore that $V(G_2) - V_2 \neq \phi$. We have already noted that for each $u \in V(G_2)$, there exists in G a set of n disjoint paths joining u and S. If $u \in V(G_2) - V_2$, then at least $n - n_2 - n_3 = n_1$ of these paths contain no vertices of $V_2 \cup V_3$. In this case, denote the set of end-vertices in S of these n_1 (or more) paths by R(u). Thus for each $u \in V(G_2) - V_2$, there exists a set $R(u) \subseteq S - V_3$ such that there are disjoint paths containing no elements of $V_2 \cup V_3$ which join u and R(u) where $|R(u)| \geq n_1$. If there exist vertices $u_1, u_2 \in V(G_2) - V_2$ such that $R(u_1) \cap R(u_2) = \phi$, then $|S - V_3| \geq 2n_1$ so that $n - n_3 \geq 2n$, and $n_2 \geq n_1$. Otherwise, let $R = \bigcup R(u)$, the union taken over all $u \in V(G_2) - V_2$, and let $G' = \langle R \cup (V(G_2) - V_2) \rangle$. It is now easy to verify that every two vertices of G' are connected so that G' itself is connected. Hence G' is a subgraph of a component of G*. Since the order of G' is a subgraph of a component of G*. Since the order of G' is at least $n_1 + (p-m-n) - n_2$, there must be a component of G* of order at most $m + n_2 - n_1$. Therefore, $m \leq m + n_2 - n_1$ so that $n_2 \geq n_1$. Thus in any case, $n_2 \geq n_1$.

The inequality $n_2 \geq n_1$ implies that $n_1 \leq n/2$. We next verify that $V(G_1) - v_1 \neq \phi$ or, equivalently, that $n_1 < m$. Assume that $n_1 = m$ so that $V(G_1) = V_1$. Hence for each $v \in V(G_1)$,

$$\deg v \leq (n_1-1) + n \leq (3n-2)/2,$$

which contradicts the fact that $\mu(G) \geq (3n-1)/2$. We conclude therefore that $n_1 < m$ and $V(G_1) - v_1 \neq \phi$.

Let $F = \langle (V(G_1) - V_1) \cup (S - V_3) \rangle$. We show that F is disconnected. Suppose, to the contrary, that F is a connected subgraph of G*. Since G* is not connected, $V(G_2) - V_2 \neq \phi$. Because each $u \in V(G_2) - V_2$ is joined to $S - V_3$ by at least n_1 paths in G*, it follows that G* is connected which is impossible. Thus F is disconnected.

Denote the components of F by F_t, t = 1, 2, ..., k, where $k \geq 2$. Furthermore, for each t = 1, 2, ..., k, denote by W_t the set of vertices of F_t in S, where $|W_t| = s_t$. We note that each $W_t \neq \phi$; for otherwise there would exist a component of F of order less than m contained in $\langle V(G_1) - V_1 \rangle$ which would also be a component of G^*.

We claim that precisely one of the subgraphs F_t contains elements of $V(G_1) - V_1$. Assume this is not the case so that there are two subgraphs F_i and F_j, $1 \neq j$, containing elements of $V(G_1) - V_1$. Let $W_i' = \cup W_t$, $t \neq i$, where $|W_i'| = s_i'$. Each of the sets $V_1 \cup V_3 \cup W_i$ and $V_1 \cup V_3 \cup W_i'$ is a cut set of G, for in each case the removal of the set from G produces a graph having a component contained in $\langle V(G_1) - V_1 \rangle$. This implies that $n_1 + n_3 + s_i \geq n$ and $n_1 + n_3 + s_i' \geq n$ so that $s_i \geq n_2$ and $s_i' \geq n_2$. However, the equality $n_1 + n_2 + n_3 = s_i + s_i' + n_3 = n$ together with the inequality $n_2 \geq n_1$ yield $s_i = s_i' = n_1 = n_2$. Therefore, $V_1 \cup V_3 \cup W_i$ is an n-cut set of G, but the graph $G - (V_1 \cup V_3 \cup W_i)$ has a component of order less than m. This produces a contradiction; henc exactly one of the subgraphs F_t contains elements of $V(G_1) - V_1$. Let F_1 be the subgraph with this property.

Now $V_1 \cup V_3 \cup W_1$ is a cut set of G so that $n_1 + n_3 + s_1 \geq n$ or $s_1 \geq n_2$. Let G_1^* be a component of G^* which contains vertices of W_1'. If $V(G_1^*) \subseteq W_1'$, then $s_1' \geq m$, but this implies that

$$n = s_1 + s_1' + n_3 \geq n_2 + m + n_3 > n_2 + n_1 + n_3 = n,$$

which is impossible. Therefore, G_1^* contains vertices of $V(G_2) - V_2$, which incidentally shows that $V(G_2) - V_2 \neq \phi$.

We show next that $V_2 \cup V_3 \cup W_1'$ is a cut set of G. Suppose this is not so. Then $G' = G - (V_2 \cup V_3 \cup W_1')$ is connected. However, $G^* = \langle V(G'') \cup W_1' \rangle$ is disconnected; therefore, G^* has a component which is a subgraph of $\langle W_1' \rangle$, but we have seen that every component of G^* which contains elements of W_1' also contains elements of $V(G_2) - V_2$. Hence $G - (V_2 \cup V_3 \cup W_1')$ is disconnected so that $V_2 \cup V_3 \cup W_1'$ is a cut set of G. This produces the inequality $n_2 + n_3 + s_1' \geq n$ or $s_1' \geq n_1$.

We now know that $s_1 + s_1' = n_1 + n_2$, $s_1 \geq n_2$, and $s_1' \geq n_1$. From this we conclude that $s_1 = n_2$ and $s_1' = n_1$. Returning to the set $V_1 \cup V_3 \cup W_1$, we note that this is an n-cut set. However, $G - (V_1 \cup V_3 \cup W_1)$ contains a component of order less than m. This produces a contradiction, and the desired result follows.

Using the construction in [8], we show that the above result is best possible, i.e., for each positive integer n and positive integer $m < (3n-1)/2$, there is a critically n-connected graph G with $\mu(G) = m$.

For $n \geq 2m + 2$, define the collection $\{H_{n,m}\}$ of graphs as follows:

$$H_{n,m} = \begin{cases} 2K_{m+1} & \text{for } n = 2m + 2 \\ K_{n-2m-2} + 2K_{m+1} & \text{for } n > 2m + 2. \end{cases}$$

It is easily seen that $H_{n,m}$ has n points and $\mu(H_{n,m}) = n - m - 2$. Using a result of [2], the equality $\varkappa(H_{n,m}) = n - 2m - 2$ follows.

For $n < m < (3n-1)/2$, define

$$G_{n,m} = H_{n,m-n} + 2K_{m-n+1} .$$

From the information obtained about $H_{n,m}$, it follows that $\mu(G_{n,m}) = m$, and, with the aid of the above mentioned result in [2], $\varkappa(G_{n,m}) = n$. The graph $G_{n,m}$ also has the property that $\varkappa(G_{n,m} - v) = n - 1$ for each point v of G.

3. <u>Minimally n-line-connected</u>. In section 2 it was pointed out that the concepts of "critically n-connectedness" and "minimally n-connectedness" are independent of each other. Analogously, the same is true of the concepts of "critically n-line-connectedness" and "minimally n-line-connectedness".

For each $n \geq 2$, the graph $K(n,n+1)$ is minimally n-line-connected, but not critically n-line-connected. We shall now present a class of graphs $\{H_n\}$ such that for each $n \geq 2$, H_n is critically n-line-connected, but not minimally n-line-connected. For $n = 2$, let $H_2 = G_1$, where G_1 is the graph in Figure 1. For $n > 2$, let $H_n = 2K_{n-1} + K_2$. It is easily seen that $\varkappa(H_n) = 2$ and $\mu(H_n) = n$. That $\lambda(H_n) = n$ follows from the result in [1] which states that if $\mu(G) \geq (p-1)/2$ for a graph with p points, then $\lambda(G) = \mu(G)$. Since H_n has 2n points and $\mu(H_n) = n > (2n-1)/2$, we have $\lambda(H_n) = \mu(H_n) = n$. The graph H_n is critically n-line-connected since the removal of any point of H_n reduces its minimum degree, and thus its line-connectivity. If the line of K_2 in H_n is labeled x, then since $\mu(H_n - x) = n$, we see as above that $\lambda(H_n - x) = n$. Hence H_n is not minimally n-line-connected.

It is easily observed that a graph is minimally 1-line-connected if, and only if, it is a tree; thus if G is a graph with $\lambda(G) = 1$ and $\mu(G) \geq 2$, then G is not minimally 1-line-connected. If $\lambda(G) = 1$, then G is connected and so there exists a point v of G such that $\lambda(G - v) = 1$. Hence there are no critically 1-line-connected graphs.

Before stating the main theorem of this section, we point out that every n-line-connected graph contains a critically n-line-connected subgraph and a minimally n-line-connected subgraph.

Theorem 2. There are no minimally n-line-connected graphs with $\mu(G) > n$.

Because the proof of this theorem is a dual of the proof of Theorem 1, it is not included here. (A proof can be found in [9]). We do, however, show that the result is the best possible. This follows from the fact that $\lambda(K_{n+1}) = \mu(K_{n+1}) = n$ and K_{n+1} is minimally n-line-connected. Another example is that of any regular graph G of degree n such that $\lambda(G) = n$.

We conclude this section by presenting a relation between "point critical graphs" and "line critical graphs", and a relation between "point minimal graphs" and "line minimal graphs".

Theorem 3. Let $\varkappa(G) = \lambda(G)$.
    (1) If G is critically n-line-connected, then G is critically n-connected.
    (2) If G is minimally n-line-connected, then G is minimally n-connected.

Proof. We only prove part of (1), the proof of part (2) being similar. Since G is critically n-line-connected, $\lambda(G-v) = n-1$ for each point v of G. Now $\varkappa(G-v) \leq \lambda(G-v) = n-1$. The removal of a single point from a graph can reduce its connectivity by at most one, thus $\varkappa(G-v) = n-1$ and G is critically n-connected.

REFERENCES

1. G. Chartrand, A graph-theoretic approach to a communications problem, J. SIAM Appl. Math., 14 (1966), 118-181.

2. G. Chartrand and F. Harary, Graphs with prescribed connectivites, Theory of Graphs, Proceedings of the Colloquium Held at Tihany Hungary, 1968, 61-63.

3. G. Chartrand, A. Kaugars, and D.R. Lick, Critically n-connected graphs, submitted for publication.

4. G.A. Dirac, Minimally 2-connected graphs, J. Reine Angew. Math., 228 (1967), 204-216.

5. F. Harary, Graph Theory, Addison-Wesley, 1969.

6. R. Halin, A theorem on n-connected graphs, J. Combinatorial Theory, to appear.

7. A. Kaugars, A Theorem on the Removal of Vertices from Blocks, Senior Thesis, Kalamazoo College, 1968.

8. D.R. Lick, Connectivity preserving subgraphs, Mathematical Report No. 1, Western Michigan University, July 1968.

9. D.R. Lick, Edge connectivity preserving subgraphs, <u>Mathematical</u>
    <u>Report No. 7</u>, Western Michigan University, September 1968.

10. M.D. Plummer, On minimal blocks, <u>Trans. Amer. Math. Soc.</u>, 134
    (1968), 85-94.

11. H. Whitney, Congruent graphs and the connectivity of graphs,
    <u>Amer. J. Math</u>., 54 (1932), 150-168.

ON RECONSTRUCTION OF GRAPHS[1]

Bennet Manvel, University of Michigan

In [15] Ulam proposed the following conjecture:

__Ulam's Conjecture__: Any graph G with $p \geq 3$ points is uniquely determined by its subgraphs $G_i = G - v_i$, $i = 1, \ldots, p$.

Harary [6] suggested that it might be best to view this problem as one of reconstructing the graph from its collection of subgraphs. This could be done uniquely only if the conjecture were true. We present in this paper a summary of progress on this problem, as well as on some related ones involving four different strengthenings of the original conjecture. Definitions of all terms used here may be found in [5].

__Progress on Ulam's Conjecture.__ The first published work on the conjecture was done by Kelly [10] who observed that disconnected or complement disconnected graphs are reconstructable. A proof is given in Harary [6]. Both of these investigators also verified the conjecture by exhaustion for all graphs with at most seven points.

Harary showed how certain invariants, including the number of points, lines, blocks, components, cutpoints, and independent cycles, as well as the degree sequence and connectivity of G, can be derived from the G_i. To display just how special the techniques involved sometimes are, we will show how the number of bridges of G may be found.

We assume that we already know the connectivity and degree sequence of G. If the connectivity is two or more, then G cannot have cutpoints, and hence has no bridges. If, on the other hand, G is disconnected then it is reconstructable so we certainly can see how many bridges it contains. Thus we are only concerned with 1-connected graphs. Furthermore, as mentioned below, Bondy has shown that any 1-connected graph without endpoints is reconstructable, so we may assume that G is a 1-connected graph with endpoints. For such a graph G with q lines, any subgraph G_i with $q - 1$ lines must have resulted from the deletion of an endpoint. Clearly then if B_i is the number of bridges in G_i there must be just $B_i + 1$ bridges in G, so we can find the number of bridges of G in every case.

Most of the common invariants of a graph which are not mentioned above have not been derived from the G_i. In particular, the problem of finding the chromatic number of G is closely related to a difficult and still unsolved extremal problem (see Dirac [3]), and the various topological invariants such as genus, thickness, and coarseness also present real difficulties.

---

[1] Research supported in part by a grant from the U.S. Air Force.

Several special kinds of graphs have been reconstructed by various investigators. Kelly [10] showed that trees are determined by their subgraphs G_i. Recently I have done the somewhat similar cases of unicyclic graphs [11] and cacti [13], the latter with D. P. Geller. Finally, Bondy [2] has shown that any 1-connected graphs without end-points is reconstructable. Since this list is, to my knowledge, exhaustive, there is almost no progress to report on reconstruction of graphs without cutpoints, except for some very special classes such as cycles, regular graphs, and eulerian graphs, which yield to elementary arguments.

The related conjecture that a graph can be reconstructed from its collection of line-deleted subgraphs was suggested by Harary [6]. Hemminger [9] pointed out that this is a special case of Ulam's Conjecture, since reconstruction of a graph G from its line-deleted subgraphs is equivalent to reconstruction of the line graph $L(G)$ from point-deleted subgraphs. For directed graphs, Harary and Palmer [7] have proved that non-strong tournaments with at least five points are reconstructable from point deletions, and any tournament is reconstructable from its line deletions. The following examples due to L. W. Beineke and E. M. Parker, respectively, show that there are strong tournaments with 5 and 6 points which are not reconstructable from their point-deleted subgraphs.

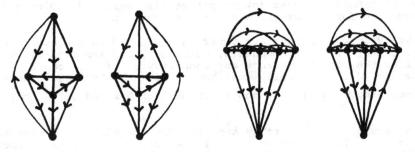

Figure 1

Strong Forms of the Conjecture. Although Ulam's Conjecture appears to be very difficult to prove, empirical evidence indicated that it is not only true but also would remain true after being significantly strengthened. I will suggest several ways this might be done, and give some evidence supporting each.

The most obvious way to make the conjecture stronger is to use fewer than p of the subgraphs G_i to reconstruct G. Thus we have the following statement.

Manvel's Conjecture. A graph G with $p \geq 3$ points is determined by any $\zeta(p)$ of its subgraphs G_i, with $\zeta(p) \leq p$, all p.

This is a strengthening of Ulam's Conjecture only if $\zeta(p) < p$, some p. Using a Fortran program on an IBM 1800, Dennis Geller and I have found that $\zeta(p) = p$ for $p = 3$, 4, and 5, but $\zeta(6) = 5$ and $\zeta(7) = 6$, suggesting that $\zeta(p) < p$, $p > 5$. The following proposition prevented the derivation of these values from being purely exhaustive.

<u>Proposition 1</u>. If two graphs G and G', with q and q' lines respectively, have p - 1 common maximal induced subgraphs, then $|q - q'| \leq 1$.

<u>Proof</u>. Let q_i be the number of lines in $G_i = G - v_i$, d_i be the degree of v_i, and q_i', v_i' be the corresponding numbers for G_i'. If we suppose that $G_i = G_i'$, i = 1, 2, ..., p-1, then clearly $q_i = q_i'$ for those i. Furthermore, since $q_i = q - d_i$ and $q_i' = q' - d_i'$ for all i we must have

(1) $$d_i' = d_i - k, \quad i = 1, 2, \ldots, p-1,$$

where k = q - q'.

Now note that since $\sum_1^p d_i = 2q$ and $\sum_1^p d_i' = 2q' = 2(q-k)$, we have

$$\sum_1^p d_i' = \sum_1^{p-1} (d_i - k) + d_p' = (2q - d_p - (p-1)k) + d_p' = 2q - 2k.$$

Thus we see that

(2) $$d_p' = d_p + (p-3)k.$$

Suppose, for simplicity, that the points v_i are arranged so that $d_1 \leq d_2 \leq \ldots \leq d_{p-1}$. Then if k is 2 or more, it is clear by (1) that the isomorphism between G_1 and G_1' must send v_2 to v_p'. Thus $|d_2 - d_p'| \leq 1$, and we see, using (2), that

(3) $$|d_2 - (d_p + (p-3)k)| \leq 1.$$

From this we conclude that $(p-3)k - 1 \leq d_2 - d_p$. Since $d_2 - d_p \leq p - 1$, this implies

(4) $$(p-3)k \leq p - 1$$

which is false for $p \geq 6$. Since it is easy to check that k is at most 1 for $p \leq 5$, the proposition is proved.

The following class of simple examples which show that $\zeta(p) \leq [p/2] + 2$ is from the paper by Manvel and Stockmeyer [14]. No better lower bound is known.

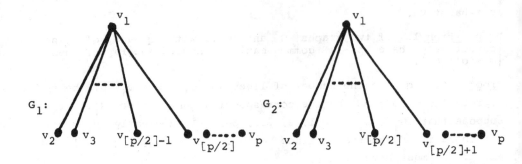

Figure 2

Another way to strengthen Ulam's Conjecture applies only to certain graphs, namely those with endpoints. Bondy [2] has made the following conjecture in which k remains to be determined.

<u>Bondy's Conjecture</u>. Any graph G with at least k endpoints can be reconstructed from those subgraphs $G_i = G - v_i$ for which v_i is an endpoint.

The graphs shown in Figure 3 were found by P.M. Neumann. They appear in Bondy's paper to show that k must be at least 4. Apparently Neumann also has an example showing k = 4 is still not enough. No one, least of all Bondy, expects that this conjecture is true for any k, but it seems very difficult to find counterexamples for large k.

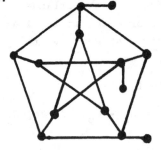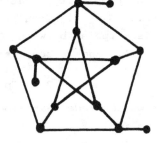

Figure 3

Harary and Palmer [8] showed that Bondy's Conjecture (with k = 2) is true for trees, and Bondy [1] improved their result by using only the peripheral endpoints. Bondy [2] and Greenwell and Hemminger [4] have recently proved the conjecture for some other classes of 1-connected graphs with endpoints.

Kelly apparently laid to rest still another strong form of Ulam's Conjecture when he stated in his pioneering paper [10] that graphs are not determined by their collection of p - 2 point subgraphs. It later developed that he had found graphs with four and five points which are not so determined, and had concluded this was probably the general situation. It turns out that every graph with six points is so determined, and no one has found an infinite class of counterexamples to the following conjecture for any number n. We have named this conjecture after Kelly since he originally raised the possibility of deleting several points.

<u>Kelly's Conjecture</u>. For any positive integer n there exists a number $\nu(n)$ such that any graph with at least $\nu(n)$ points is uniquely determined by its collection of n-point deleted subgraphs.

The empirical data mentioned above suggests that $\nu(2) = 6$, while Ulam's Conjecture claims that $\nu(1) = 3$. It is unfortuate that it would be so difficult to obtain even a guess for $\nu(3)$. The only general progress made on deletion of several points is reported to me by Joel Spencer of Rand who says he has reconstructed trees from their collection of subgraphs T_{ij} for which both T_i and T_{ij} are trees.

Most of the invariants which can be found from a collection of 1-point deleted subgraphs are extremely elusive in the n-point deleted case, except under certain restrictions. One exception is q, the number of lines of G. If n points are deleted at a time and $q_{i_1 i_2 \ldots i_n}$ is the number of lines in one subgraph, then it is clear that

(5)
$$q = \frac{\sum q_{i_1 \ldots i_n}}{\binom{p-2}{n}} .$$

Thus q may be found as long as p-n is at least 2. On the other hand, the degree sequence of G is difficult to find unless the maximum degree of G is no more than p - n - 1. In general, the whole situation is much more complex than the 1-point deleted case.

Our final conjecture is due to Harary [6]. Most people tend to find it very unlikely, but no one has a counterexample, or even an idea of where to look for one.

<u>Harary's Conjecture</u>. A graph G with $p \geq 4$ points can be reconstructed uniquely from its set of non-isomorphic subgraphs $G_i = G - v_i$.

Note that we know here only what graphs are in the list of subgraphs G_i, not how many times each occurs there. I have shown [12] that this conjecture is true for trees, and, in fact, trees are determined by their sets of nonisomorphic subtrees (i.e., only those subgraphs resulting from endpoint deletions), except in the two cases shown in Figure 4.

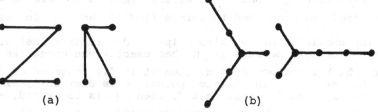

(a)   (b)

Figure 4

It appears that Bondy's reconstruction of 1-connected graphs without endpoints can be modified to show that Harary's Conjecture holds for such graphs, but the details must still be worked out.

The derivation of invariants for G from the information given in Harary's Conjecture is not as neat as one would hope, since it usually involves the treatment of several possible cases. As an example, we present the next result.

Proposition 2. The number of lines of a graph G with at least 4 points can be found from its set of non-isomorphic subgraphs G_i.

Proof. If q is the number of lines in G, q_i the number in G_i, and δ the minimum degree among the points of G, then it is clear that

$$(6) \qquad\qquad q = \delta + \max_i q_i .$$

Thus to find q we need merely find δ. In general δ is just one greater than the minimum degree found in any of the G_i. The exception is, of course, when $\delta = 0$. Thus we must distinguish whether or not G has any isolated points.

If each G_i has an isolated point, then G has also unless it is just a union of copies of K_2, the complete graph on 2 points, which can easily be recognized. On the other hand, if at least two of the G_i have no isolated point, then G clearly cannot have one. The only difficult case is when G has several G_i's and exactly one of them, say G_1, has no isolated point. If this is the case then in general G has an isolated point, namely v_i. If it does not, then any two points v_i, v_j whose deletion results in a subgraph without an isolated point must be such that $G - v_i \cong G - v_j \cong G_1$. If we call the set of such points V_1, there are two possibilities for G. It is either a point-symmetric graph on V_1 together with some copies of K_2 on $V - V_1$, or else it is a point-symmetric graph on $V - V_1$ plus k (>0) endlines on each point, terminating at the points of V_1. In either case G has just two G_i's, neither of which is point-symmetric (unless G is $K_2 \cup K_3$, which can be recognized). But a graph with an isolated point v_1 and just two G_i's must have G_1 point-symmetric. Thus we can decide when G has an isolated point and therefore find δ and q in every case.

The knowledge of δ also helps in determining whether G is connected or not, because if G has exactly one connected G_i and others which are disconnected, then it is disconnected or connected as it has or hasn't an isolated point. More obviously, if there is just one G_i and it is connected, then G is connected, and similarly if there are two different connected G_i's. Finally, if all of the G_i are disconnected, then G is also. Thus we may always decide whether G is or is not connected. It is easy to see that Harary's Conjecture holds for disconnected and complement-disconnected graphs.

Although it is very unlikely that anyone will soon prove any of these conjectures, it is possible that someone investigating them may have new insights into Ulam's Conjecture itself. On the other hand, those looking for counterexamples to the original conjecture should find it easier to first produce them for these statements.

REFERENCES

1. J.A. Bondy, On Kelly's congruence theorem for trees. Proc. Cambridge Phil. Soc. (to appear)

2. J.A. Bondy, On Ulam's Conjecture for separable graphs. Canad. J. Math. (to appear)

3. G.A. Dirac, A property of 4-chromatic graphs and some remarks on critical graphs. J. London Math. Soc. 27 (1952), 85-92.

4. D.L. Greenwell and R.L. Hemminger, Reconstruction graphs (this volume).

5. F. Harary, Graph theory. Addison-Wesley, Reading, 1969.

6. F. Harary, On the reconstruction of a graph from a collection of subgraphs. Theory of graphs and its applications (M. Fiedler, ed.), Prague, 1964, pp. 47-52; reprinted Academic Press, New York, 1964.

7. F. Harary and E.M. Palmer, On the problem of reconstructing a tournament from subtournaments. Monat. für Mathematik 71 (1967), 14-23.

8. F. Harary and E.M. Palmer, The reconstruction of a tree from its maximal proper subtrees. Canad. J. Math. 18 (1966), 803-810.

9. R.L. Hemminger, On reconstructing a graph. Proc. Amer. Math. Soc. 20 (1969), 185-187.

10. P.J. Kelly, A congruence theorem for trees. Pacific J. Math. 7 (1957), 961-968.

11. B. Manvel, Reconstruction of unicyclic graphs. Proof techniques in graph theory (F. Harary, ed.) Academic Press, New York, 1969.

12. B. Manvel, Reconstruction of trees. Canad. J. Math. (to appear).

13. B. Manvel and D.P. Geller, Reconstruction of cacti. Canad. J. Math. (to appear).

14. B. Manvel and P.K. Stockmeyer, On reconstruction of matrices. (submitted for publication).

15. S.M. Ulam, <u>A collection of mathematical problems</u>. Wiley (inter-
    science), New York, 1960, p. 29.

THE COHESIVE STRENGTH OF GRAPHS

David W. Matula, Washington University

1. Introduction and Summary. Many graph theoretic problems can be
decomposed and simplified by treating the problem separately on the
components and/or blocks of the graph, and it is of interest to consider other subgraphs of a graph characterized by their connectivities
over which certain graph theoretic questions may be decomposed and re-
solved. It is generally geometrically evident that the intensity of
connectivity can vary widely over different parts of a graph, and in
order to determine special subgraphs characterized by their connectivities it is expedient to have a measure of the local intensity of connectivity within a graph.

In this paper the edge connectivity of subgraphs is shown to be a
useful tool for characterizing the notion of the intensity of local
connectivity. First the maximal k-edge connected subgraphs (herein
termed k-components) are considered and shown to retain the important
property of being vertex disjoint. The cohesiveness function, defined
for each vertex and edge to be the maximum edge connectivity of any
subgraph containing that element, is then introduced and shown to be a
useful tool for determining the k-components for all k.

A sequence of cuts which separate the graph into isolated vertices is termed a slicing, and a natural dual relation between cohesiveness and slicings is exposed. The maximum number of edges in any cut
of a slicing is called the width of the slicing, and our main result
then appears as a fundamental min-max theorem between the slicings and
subgraphs of a graph: The minimum width of any slicing is equal to
the maximum edge connectivity of any subgraph. A slicing which util-
izes a minimum cut at every step is termed a narrow slicing and is
shown to be instrumental in computing the cohesiveness function for all
elements of a graph.

Finally, as an application of the use of the cohesiveness function in another area of graph theory, the result is presented that the
maximum value of the edge connectivity over all subgraphs (i.e. the
maximum cohesiveness) plus unity is an upper bound on the chromatic
number of a graph.

2. The Edge Connectivity of Graphs and Subgraphs. A cut set or cut
$C = (A, \overline{A})$ of G is an edge set composed of all edges of G with
one endpoint in the non void proper subset A of V(G) and the other
endpoint in $\overline{A} = V(G) - A$. The cut $(A, \overline{A})$ is said to separate the
vertices of A from the vertices of $\overline{A}$, so that

$$G - (A, \overline{A}) = \langle A \rangle \cup \langle \overline{A} \rangle \qquad (1)$$

A connected graph with at least two vertices in which every cut
set has at least k edges, $k \geq 1$, is k-edge connected, and the
graph consisting of a single vertex is said to be k-edge connected
only for k = 1. For any $k \geq 1$, a k-component of G is a maximal

k-edge connected subgraph of G.   Note that the l-components of G are
precisely the components of G, and the notion of k-components for
$k \geq 2$ represents an extension of the concept of component which relates to more tightly knit subgraphs.

The term "k-edge component" would be more definitive than k-component since separating sets of vertices have also been utilized to
generalize the notion of component (see [3]); however, in this paper
we shall deal exclusively with connectivity through edges so that no
confusion should occur with the abbreviated term k-component.

For any graph G with at least two vertices the edge connectivity, $\lambda(G)$, is the minimum number of edges in any cut of G, i.e.

$$\lambda(G) \equiv \min \left\{ |C| \mid C \text{ is a cut of } G \right\} \qquad (2)$$

and any cut of G with $\lambda(G)$ edges is a minimum cut of G. The
graph consisting of a single vertex has no cut sets and the edge connectivity in this case is taken to be unity, so that $\lambda(G) = 0$ if and
only if G is not connected. Hence for any $k \geq 1$, G is k-edge
connected if and only if $k \leq \lambda(G)$, and any k-components of G must
have edge connectivity at least k. Furthermore, any subgraph, G',
of G must be contained in some $\lambda(G')$-component of G.

The analysis of selected subgraphs with regards to their edge
connectivity has interest both of itself and as a tool in attacking
other and quite diverse graph theoretic questions. Certainly the implicit property of disjointedness of the ordinary components of a
graph provides an elegantly simple and valuable tool to the researcher,
as evidenced by the fact that many graph theoretic proofs are composed
by looking at the given problem separately on each component. It will
now be shown that this important disjointedness property of components
carries over in our edge connectivity based generalization to k-components, which is all the more significant since the alternative generalization to a k-(vertex) component based on vertex connectivity does
not inherit the disjointedness property for any $k \geq 2$. Actually a
stronger result will be demonstrated from which the fact that maximal
k-edge connected subgraphs are vertex disjoint follows immediately.

Lemma. Let $G_1, G_2, \ldots, G_n$ be subgraphs of G such that $\bigcup_{i=1}^{n} G_i$
is connected. Then

$$\lambda(\bigcup_{i=1}^{n} G_i) \geq \min_{1 \leq i \leq n} \{\lambda(G_i)\} . \qquad (3)$$

Proof. If $\bigcup_{i=1}^{n} G_i$ consists of a single vertex, the result is immediate. Otherwise let $C = (A, \overline{A})$ be a minimum cut of $\bigcup_{i=1}^{n} G_i$, which
must contain at least one edge since $\bigcup_{i=1}^{n} G_i$ is connected and has at
least two vertices. If C contains an edge of G_j, both A and $\overline{A}$
contain vertices of $V(G_j)$, so that C contains a cut of G_j. Hence

$$\lambda(\bigcup_{i=1}^{n} G_i) \geq \lambda(G_j) \quad \text{for at least one} \quad G_j, \quad 1 \leq j \leq n, \quad \text{proving the lemma.}$$

<u>Corollary</u>. For any $k \geq 1$, the k-components of a graph are vertex disjoint.

It is evident that the vertices and edges of a graph participate to varying extents in the formation of tightly knit substructures within the graph, and from the preceding discussion it is reasonable to suggest the edge connectivity of subgraphs as the appropriate mathematical tool for investigating this local intensity of connectivity. To provide a numerical measure for this concept of local connectivity we introduce the cohesiveness function.

For every element $x \in V(G) \cup E(G)$ of the graph G the <u>cohesiveness</u> of x, denoted $h(x)$, is defined to be the maximum value of the edge connectivity of any subgraph of G containing x. A suscinct global measure of the cohesiveness of G is provided by the maximum value of $h(x)$ over $x \in V(G) \cup E(G)$, and this is termed the cohesive strength or simply the <u>strength</u>, $\sigma(G)$, of G. Thus

$$\sigma(G) \equiv \max \{\lambda(G') \mid G' \text{ is a subgraph of } G\}. \tag{4}$$

The cohesiveness function can be represented by the (symmetric) <u>cohesiveness matrix</u> whose columns and rows are the vertices of G, and the entry in the v_i, v_j position is either the cohesiveness of the edge $\{v_i, v_j\}$ or zero if $\{v_i, v_j\}$ is not an edge of G. The cohesiveness of a vertex v_i is then the maximum element in the row corresponding to v_i, and the strength, $\sigma(G)$, is the maximum term in the cohesiveness matrix.

Clearly $h(x) \geq 1$ for all $x \in V(G) \cup E(G)$, and $\sigma(G) = 1$ if and only if G is a forest. Although the presence of a single cycle in G assures that $\sigma(G) \geq 2$, the fact that some edge is in more than one cycle does not insure that $\sigma(G) \geq 3$. Figure 1 exhibits a graph of strength 4 along with its cohesiveness matrix.

The cohesiveness function is of value in determining the k-components of a graph, G, for it can be shown (See [4]) that the components of the subgraph obtained by deleting all elements of G of cohesiveness less than k are precisely the k-components of G. The sets of vertices composing the k-components have been outlined in the graph of Figure 1. Note that the set boundaries then appear as contours of cohesiveness.

In summary, a numerical measure of the intensity of (edge) connectivity over the subgraphs of a graph has been realized in the cohesiveness function, and we now seek to

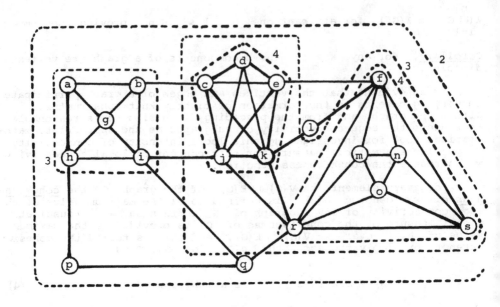

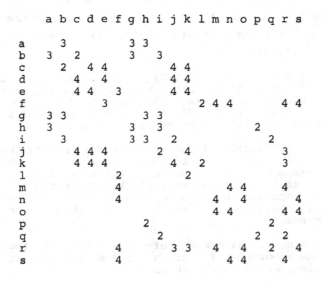

	a	b	c	d	e	f	g	h	i	j	k	l	m	n	o	p	q	r	s
a		3					3	3											
b	3		2				3		3										
c		2		4	4					4	4								
d			4		4					4	4								
e			4	4		3				4	4								
f					3							2	4	4				4	4
g	3	3						3	3										
h	3						3		3							2			
i		3					3	3		2							2		
j			4	4	4				2		4							3	
k			4	4	4					4		2						3	
l						2					2								
m						4								4	4			4	
n						4							4		4				4
o													4	4				4	4
p								2									2		
q									2							2		2	
r						4				3	3		4		4		2		4
s						4								4	4			4	

Figure 1. A graph and its cohesiveness matrix. The k-components of the graph have been outlined.

further understand the behavior of the cohesiveness function and to compute its value over the elements of a graph. Just as a cut set of a graph may be interpreted as an important notion dual to the notion of connectivity in a graph, we now introduce the concept of a slicing of a graph as a dual to the notion of cohesiveness.

The ordered partition of the edges of the graph G, $Z = (C_1, C_2, \ldots, C_m)$, is a __slicing__ of G if

$$
C_i \text{ is a cut set of }
\begin{cases}
G & \text{for } i=1 \\[2ex]
G - \bigcup\limits_{j=1}^{i-1} C_j & \text{for } 2 \leq i \leq m
\end{cases}
\tag{5}
$$

and each C_i will also be termed a __cut of the slicing__. Furthermore, Z is a __narrow slicing__ if each such cut C_i is a minimum cut of some component of $G - \bigcup\limits_{j=1}^{i-1} C_j$, where here and subsequently $G - \bigcup\limits_{j=1}^{0} C_j \equiv G$. The notion of slicing applies only to graphs with at least one edge since the void partition is not recognized.

Thus a slicing of G has a dynamic interpretation as a sequence of cuts which stepwise separate G into more and more components and finally to isolated vertices. Furthermore a narrow slicing effects this separation using only minimum cuts at each step. Since each cut of a slicing successively increases the number of components in the resulting graph, the number of cuts in a slicing is no greater than the number of vertices diminished by the number of components of G, with equality in the case of a narrow slicing.

Our main concern with slicings is not in the number of cuts but rather in the number of edges in the biggest cut of the slicing. Thus, the width $w(Z)$ of the slicing Z of G is defined by

$$
w(Z) \equiv \max \{|C| \mid C \text{ is a cut of the slicing } Z\}
\tag{6}
$$

and any cut of Z such that $|C| = w(Z)$ is a __wide cut__ of the slicing Z. Note that a minimum cut refers to a cut __of a graph__ whereas a wide cut refers to a cut __of a slicing__. The aforementioned claim of duality between slicings and cohesiveness will now be investigated, and, in addition, the computational relevance of narrow slicings to the determination of the cohesiveness function will be discussed.

3. __A Duality Theory for Graph Cohesiveness and Slicings__. The duality between slicings and cohesiveness in a graph is evident in the following min-max theorem.

__Theorem 1.__ For any graph G containing at least one edge, the minimum width of any slicing of G is equal to the strength of the graph, i.e.,

$$\min \{w(Z) \,|\, Z \text{ is a slicing of } G\} = \max \{\lambda(G') \,|\, G' \tag{7}$$
$$\text{is a subgraph of } G\}.$$

<u>Proof</u>. The theorem will be established by showing an inequality in each direction. Assume G is a graph containing at least one edge.

(i) $\min \{w(Z) \,|\, Z$ is a slicing of $G\} \geq \max \{\lambda(G') \,|\, G'$ is a subgraph of $G\}$. Let K be a subgraph of G with at least two vertices such that $\lambda(K) = \sigma(G)$. Then for any slicing Z of G there is a first cut C_i separating vertices of K. But then C_i must contain a cut of K, so that $|C_i| \geq \lambda(K) = \sigma(G)$. Hence $w(Z) \geq \sigma(G)$ for any slicing Z, proving (i).

(ii) $\min \{w(Z) \,|\, Z$ is a slicing of $G\} \leq \max \{\lambda(G') \,|\, G'$ is a subgraph of $G\}$. Let $Z^* = (C_1, C_2, \ldots, C_m)$ be a narrow slicing of G with wide cut C_i. Then if G_i is the subgraph of G cut by C_i, $\lambda(G_i) = w(Z^*)$, proving (ii) and the theorem.

Each cut of a slicing is actually a cut of some subgraph of G, the particular subgraph in question being determined by the preceding cuts as well as the current cut of the slicing. These particular subgraphs are referred to in [4] as the subgraphs cut by the slicing, and they are of both theoretical and computational interest. For example it can be show (see [4]) that the cohesiveness of any element of a graph other than an isolated vertex is equal to the largest edge connectivity of exactly those subgraphs cut by any particular narrow slicing and which contain that element. Furthermore each k-component of a graph for any and all k must be one of the subgraphs cut by any narrow slicing.

Now a narrow slicing of a graph can be computed by utilizing the techniques of network flow theory [1], in particular the cut trees [2] of each of the subgraphs cut by the narrow slicing must be systematically determined. More details are given in [4], wherein it is estimated that the growth of solution time in determining a narrow slicing of a graph G will be bounded roughly by the third power of the number of vertices in G.

The duality between slicings and cohesiveness along with the computational algorithm for determining a narrow slicing provide a neat package for analyzing the concept of the local intensity of connectivity within a graph. As an added bonus to this effort we shall close with an application of these concepts to the determination of an apparently new upper bound on the chromatic number.

4. <u>An Application to Bounding the Chromatic Number</u>. It is reasonable to expect that the intensity of connectivity within a graph should be related to the chromatic number, and in particular, the more intensely connected subgraphs, rather than merely the average intensity of connectivity over the graph, should generally pinpoint the constraining conditions determining the chromatic number.

With regards to the edge connectivity of subgraphs the former expectations are born out in the following theorem relating the chromatic number to the strength of the graph.

__Theorem 2__. Let $\chi(G)$ be the chromatic number of G. Then

$$\chi(G) \leq \sigma(G) + 1, \tag{8}$$

A proof [4] of this theorem which yields a coloring algorithm follows readily from our main theorem using a narrow slicing and making recursive use of a constructively demonstrated result of Ore [6, p. 232] which may be stated such that if $(A,\overline{A})$ is a cut of G, then $\chi(G) \leq \max \{\chi(<A>), \chi(<\overline{A}>), |(A,\overline{A})|+1\}$. This upper bound on the chromatic number appears to be better than other bounds in the literature. In particular it contains the result reported by Szekeres and Wilf [7] and this author [5] elsewhere, which states that $\chi(G)$ is bounded above by a number one greater than the maximum over the subgraphs of G of the minimum degree in the subgraph.

In contrast to this upper bound on the chromatic number provided by edge connectivity considerations, it is well known that a lower bound is provided by the largest number of vertices in any clique of the graph, so that it is evident that appropriate measures of the intensity of connectivity over the subgraphs of a graph do provide considerable insight into the value of the chromatic number.

REFERENCES

1. L. R. Ford and D. R. Fulkerson, _Flows in Networks_, Princeton University Press, Princeton, 1962.

2. R. E. Gomory and T. C. Hu, Multi-terminal network flows, _J. SIAM Appl. Math._ 9(1961), 551-570.

3. F. Harary, _Graph Theory_, Addison-Wesley, Reading, 1969.

4. D. W. Matula, The cohesive strength of graphs, Washington University, Department of Applied Mathematics and Computer Science, _Report_ AM-68-9, 1968.

5. D. W. Matula, A min-max theorem for graphs with application to graph coloring, _SIAM Review_, 10(1968), 481-482.

6. O. Ore, _Theory of Graphs_, Amer. Math. Soc. Collog. Publ. Vol. 38, Providence, 1962.

7. G. Szekeres and H. S. Wilf, An inequality for the chromatic number of a graph, _J. Combinatorial Theory_, 4(1968), 1-3.

HYPO-PROPERTIES IN GRAPHS

John Mitchem,[1] Western Michigan University

It is a common problem in graph theory to study those graphs which possess some specified property. However, there are usually varying degrees to which a graph may not have such a property. For example, a graph G may fail to possess a particular property, but every maximal subgraph of G may actually have the property. It is precisely this situation which we now consider.

A graph G is said to have <u>hypo-property</u> P if G does not have property P but G - v has property P for every vertex v ∈ V(G). This concept is clearly related ot the familiar concept of "critical". A graph G is <u>critical with respect to property</u> Q if G possesses property Q but G - v for every v ∈ V(G) does not have property Q. Thus, if ~P denotes the negation of P, then G has hypo-property P if and only if G is critical with respect to property ~P. In an analogous manner, one can relate the edge counterparts: hyper-properties versus minimal graphs.

Undoubtedly, the most well-known concept to which the term "critical" has been applied is the n-chromatic graphs. A graph is n-chromatic critical if $\chi(G) = n$ and $\chi(G - v) \neq n$ for all $v \in V(G)$. Hence this is equivalent to the concept of hypo (n-1)-chromatic graphs, although the latter is clearly more cumbersome with which to deal. With some properties, however, it is more natural to consider the "hypo" point of view rather than the "critical" approach.

The hypo-hamiltonian graphs have been the subject of several recent studies. In particular, it has been shown in [3] that the smallest hypo-hamiltonian graph has order 10 and is, in fact, unique. (It is probably not surprising that this graph turned out to be the Petersen graph, which has proved to have several interesting properties.) It has further been shown that no hypo-hamiltonian graphs of order 11 or 12 exist, while there are such graphs with 13 and 15 vertices (including a non-regular one of order 13). It has been conjectured [2] that no hypo-hamiltonian graph has girth less than five. Of more importance, however, is the result, obtained independently by Sousselier (see [2]) and Lindgren [5] that for all n ≥ 1, there exist hypo-hamiltonian graphs of order 6n + 4. A graph with a hamiltonian path is called traceable. It has been conjectured [4] that no graph is hypo-traceable.

Closer to our own interest is the work of Wagner [6] who investigated and classified the hypo-planar graphs (or equivalently the critically non-planar graphs). Wagner proved that the class of all hypo-planar graphs can be divided into four subclasses. Among the

[1]Research supported in part by a Faculty Research Grant from Western Michigan University. This work constitutes part of a doctoral thesis to be presented to WMU.

graphs which are hypo-planar are graphs homeomorphic from K_5, certain non-planar graphs which are embeddable in the Möbius strip so that the border of the Möbius strip is a Hamiltonian cycle, and various non-planar graphs which consist of a path P, a cycle disjoint from P, and edges joining endpoints of the path with points on the cycle.

With these general remarks and historical comments in mind, we now turn our attention to the main topic of this lecture, namely a special class of planar graphs. A graph G is outerplanar if G can be embedded in the plane in such a way that all vertices of G lie on the exterior region. We present a characterization of hypo-outerplanar graphs.

First from the well known facts that every outerplanar graph with $p \geq 2$ points and q lines has at least two points of degree two or less and $q \leq 2p - 3$ we can establish the following:

Proposition 1. Any hypo-outerplanar graph has at least two points of degree not exceeding 3.

Proposition 2. If G is hypo-outerplanar with p points and q lines, then $q \leq 2p - 2$.

Proof. Let G be a hypo-outerplanar graph with p points and q lines and for any vertex v of G, let q_v be the number of lines in $G - v$. Since $G - v$ is outerplanar, $q_v \leq 2(p - 1) - 3 = 2p - 5$ for all points v in G. Proposition 1 implies that G has a point, call it u, of degree not more than 3. But $q_u \leq 2p - 5$ and thus $q \leq q_u + 3 \leq 2p - 2$.

We observe that for hypo-outerplanar graph K_4 $q = 2p - 2$.

Before stating our characterization of hypo-outerplanar graphs, we define a few basic terms.

A graph G homeomorphic from $K_{2,3}$ is called a theta-graph. The two points in G of degree 3 are called primary points and are denoted v_1 and v_2. A point u adjacent to a primary point v_i is called a secondary point and we say u is secondary to v_i.

A super theta-graph is a theta graph together with exactly one of the following:
   i)   a line adjacent to the primary points
   ii) a line joining two points secondary to v_1 and/or a line joining two points secondary to v_2.

A graph G is called a quasi-wheel if G is a cycle together with a new point u adjacent with at least three points of the cycle. The point u is called the hub of G.

We are now ready to commence the characterization of hypo-outerplanar graphs. Chartrand and Harary [1] have shown that a graph is outerplanar if and only if it has no subgraph homeomorphic from K_4

or $K_{2,3}$. Thus theta graphs, super theta graphs, and quasi-wheels are not outerplanar. By routine checking we obtain

Proposition 3. If graph G is a theta graph, a super theta-graph or a quasi-wheel, G is hypo-outerplanar.

Before continuing our characterization of hypo-outerplanar graphs we need a number of lemmas.

Lemma 1. If graph G has a subgraph H which is homeomorphic from K_4 or $K_{2,3}$ and $V(G) \neq V(H)$, then G is not hypo-outerplanar.

Proof. This lemma follows immediately from the observation that H is a non-outerplanar subgraph of $G - v$ for some v in G.

Lemma 2. Let G be homeomorphic from K_4. If there are two non-adjacent lines u_1u_2 and u_3u_4 of K_4 which are subdivided to obtain G, then G is not hypo-outerplanar.

Proof. Let u be the point introduced by the subdivision of u_1u_2. Then $G - u$ has a subgraph homeomorphic from $K_{2,3}$ and hence is not outerplanar.

We need a definition before proving the final lemma. Let G be a graph which consists of a theta graph H together with extra lines. A line uv of G which is not in H which satisfies (i), (ii), or (iii) is called a _basic line_:
    i) One of the points, say u, is neither a primary nor a secondary point of H, and point v is not secondary to both primary points of H.
    ii) Point u is secondary only to v_1, and v is secondary only to v_2.
    iii) Exactly one of u and v, say v, is primary.

Lemma 3. If graph G consists of a theta-graph H together with extra lines including a basic line, then G is not hypo-outerplanar.

Proof. Let uv be a basic line of G, with u and v as given in the definition, and let P_u be the path in theta-graph H between primary points which includes u. Point v might lie on P_u or on another path in H joining primary points, or v might be a primary point. (Figures 1a, 1b, 1c.) In any case there is a point u' on P_u between u and one primary point such that $G - u'$ is not outerplanar. Hence G is not hypo-outerplanar.

We now continue our characterization of hypo-outerplanar graphs. Since such graphs are not outerplanar, we know that any graph which is hypo-outerplanar has a subgraph homeomorphic from K_4 or $K_{2,3}$.

Proposition 4. If G is hypo-outerplanar and has a subgraph homeomorphic from $K_{2,3}$, then G is a theta-graph, a quasi-wheel, or a super theta-graph.

Proof. Let G be hypo-outerplanar and have a subgraph H which is homeomorphic from $K_{2,3}$. Lemma 1 implies that $V(H) = V(G)$. If G is itself homeomorphic from $K_{2,3}$, then G is a theta-graph. Thus we need only consider a graph G which has a subgraph H homeomorphic from $K_{2,3}$, where $V(H) = V(G)$, and $H \neq G$. (i.e., G consists on H together with some extra lines). We let v_1 and v_2 be the primary points of H. Lemma 3 implies that G has no basic lines.

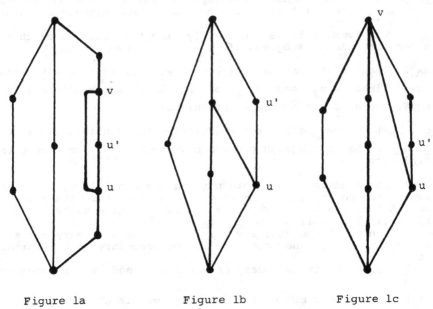

| Figure 1a | Figure 1b | Figure 1c |

We observe that non-basic lines are those incident with 2 points secondary to the same primary point, those joining primary points, and those joining a non-secondary non-primary point with a point secondary to both primary points. A number of different cases depending on the type of lines added to H are now considered.

Case i. The extra lines of H include v_1v_2.

If v_1v_2 is the only extra line, Proposition 4 implies G is hypo-outerplanar. If there are other extra lines, let u_1u_2 be one such where u_1 is secondary to v_1. Then the path in H from v_1 to v_2 which contains neither u_1 nor u_2 has a point w distinct from v_1 and v_2. G - w is not outerplanar, which contradicts our hypothesis that G is hypo-outerplanar.

Case ii. Each extra line x_i of H joins a non-secondary, non-primary point with a point which is secondary to both primary points.

In this case, if u_i is the same point for all extra lines, then G is a quasi-wheel. If $u_i \neq u_j$ for some $i \neq j$, then $G - v_1$ is not outerplanar.

Case iii. Each extra line in H joins secondary points.

If each extra line is incident with a point which is secondary to both primary points, G is a quasi-wheel. So let us suppose otherwise. Then if there are three or more lines the removal of one of the primary points leaves a non-outerplanar graph. If there are two lines which join points secondary to different primary points or just one such extra line, G is a super theta graph. If there are two extra lines joining points secondary to v_1, then $G - v_1$ is not outerplanar.

Case iv. There is at least one extra line in H which joins secondary points and at least one which joins a non-secondary, non-primary point with a point u_1 secondary to v_1 and v_2.

In this case, if each extra line is incident with u_1, G is a quasi-wheel. Otherwise there is a line $s_1 s_2$ in G with $s_1 \neq u_1 \neq s_2$ where s_1 and s_2 are secondary to v_1 and $G - v_1$ is not outerplanar.

This concludes all cases and completes the proof.

Proposition 5. If G is hypo-outerplanar and has a subgraph homeomorphic from K_4, then G is a quasi-wheel or a super theta-graph.

Proof. Let G be hypo-outerplanar and have a subgraph H which is homeomorphic from K_4. Lemma 1 implies $V(H) = V(G)$. Either G is itself homeomorphic from K_4 or G has a subgraph $H \neq G$ which is homeomorphic from K_4, and has the same point set as G. Either way, G consists of H homeomorphic from K_4, together with a set (possibly empty) of extra lines.

Thus H can be formed by a sequence of subdivisions of lines of K_4. If the sequence is empty, H is necessarily the same as G and G is a quasi-wheel. Hence, assume that the sequence of subdivisions is non-empty. If the subdivisions occur on 4 or more lines of K_4, Lemma 2 implies that H is not hypo-outerplanar. Therefore, the subdivisions can occur on at most 3 lines of K_4. Now we consider two cases.

Case i. There is a point v_4 of K_4 which is not incident with any line of K_4 which must be subdivided to form H.

In this case since H is not isomorphic to K_4, H must be obtainable from K_4 by at least one subdivision of a line not incident with v_4. Let this line be $v_1 v_2$ and label the other point of

K_4 by v_3. Let the graph formed by a single subdivision of line v_1v_2 be labeled H', and observe that $H' - (v_3v_4)$ is a copy of $K_{2,3}$. H can now be formed by a sequence (perhaps empty) of subdivisions of H'. By hypothesis, line v_3v_4 is not subdivided to obtain H, so that $H - (v_3v_4)$ is a theta-graph. Thus H consists of a graph homeomorphic from $K_{2,3}$ together with an extra line. But G is H together with a set (perhaps empty) of extra lines. Hence G consists of $H - (v_3v_4)$ which is homeomorphic from $K_{2,3}$ along with additional lines. From Proposition 4 we know all such hypo-outerplanar graphs are quasi-wheels or super theta-graphs.

Case ii. Every point of K_4 is incident with at least one line which must be subdivided to obtain H.

In this case if every point of K_4 is incident with at least one line which is subdivided and at least one line which is not subdivided to obtain H, then Lemma 2 shows that H is not hypo-outerplanar which therefore implies that G is not hypo-outerplanar.

We thus know that all lines incident with one point, say v_1, of K_4 must be subdivided in order to obtain H. Label the other points of K_4 by v_2, v_3 and v_4. $H - (v_3v_4)$ is then homeomorphic from $K_{2,3}$ because it consists of 3 disjoint paths of length two or more from v_1 to v_2. Thus G is hypo-outerplanar and consists of a theta-graph together with extra lines. Proposition 4 implies that such graphs are quasi-wheels or super theta-graphs.

This concludes the proof of Proposition 5.

Combining the first Propositions 3, 4 and 5 we obtain the following characterization of hypo-outerplanar graphs.

Proposition 6. A graph G is hypo-outerplanar if and only if G is a theta-graph, a super theta-graph, or a quasi-wheel.

This characterization yields a number of corollaries.

Corollary 1. Any hypo-outerplanar graph has at most two points of degree greater than 3 and at most one point of degree greater than 4.

Corollary 2. Hypo-outerplanar graphs are planar.

Figure 2 shows that the converse of Corollary 2 does not hold.

Before considering the final corollary we define a wheel as a quasi-wheel with p points and maximum degree equal to $p - 1$.

Wagner [6] proved that every hypo-planar graph G is 5-colorable and that G has chromatic number 5 if and only if G consists of K_2 and a cycle C_n, disjoint from K_2, where n is odd, such that every point of K_2 is adjacent to every point of C_n. We now present an analogous result for hypo-outerplanar graphs.

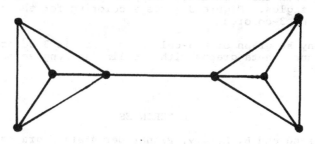

Figure 2

Corollary 3. If G is hypo-outerplanar, the chromatic number of G does not exceed 4 and G has chromatic number 4 if and only if G is a wheel with an even number of points.

Proof. Observe that a wheel is a cycle together with one point u not on the cycle such that v is adjacent with every point of the cycle. If wheel G has an even number of points, the cycle has an odd number of points and hence any coloring of G can be cone with 4 colors. It is apparent that any coloring of G requires 4 colors. If wheel G has an odd number of points, then G can be colored with three colors.

A quasi-wheel which is not a wheel can be 3-colored by coloring the hub u and one point non-adjacent to u with color 1 and then alternating colors 2 and 3 among the other points.

A super theta-graph G with line $(v_1 v_2)$ can be 3-colored by coloring v_1 with color 1, v_2 with color 2, the three points secondary with v_2 with color 3, and alternating colors 2 and 3 among the remaining points on each path in the theta-subgraph of G.

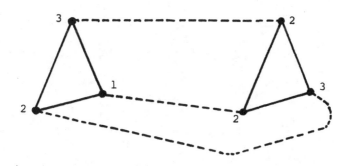

Figure 3

If G is a super theta-graph with two lines joining secondary points and if G is not a quasi-wheel, then G consists of two triangles along with three mutually disjoint paths joining points in different triangles. Figure 3 gives a coloring for the triangles so that G can be 3-colored.

Since any subgraph of a 3-colorable graph is 3-colorable, theta graphs and super theta-graphs with one line joining secondary points are 3-colorable.

REFERENCES

1. G. Chartrand and F. Harary, Planar permutation graphs, _Annales de l'Institute Henri Poincare_', 3 (1967), 433-438.

2. J.C. Herz, J.J. Duby, and F. Vigue, Recherche systematique des graphs hypo-hamiltonians, _Théorie des Graphs, Journées Internationales D'Etudes_. (Dunod, Paris, 1967), 153-159.

3. J.C. Herz, T. Gordon, and P. Rossi, Solution of probléme No. 29, _Rev. Francaise Rech. Operationelle_, 8(2), (1964), 214-218.

4. S.F. Kapoor, H.V. Kronk, and D.R. Lick, On detours in graphs, _Canad. Math. Bull._, 11 (1968), 195-201.

5. W.F. Lindgren, An infinite class of hypohamiltonian graphs, _Amer. Math. Monthly_, 74 (1967), 1087-1089.

6. K. Wagner, Fastplättbare Graphen, _J. Combinatorial Theory_, 3 (1967), 326-365.

AN EXTENSION OF GRAPHS

Douglas W. Nance, Central Michigan University

1. <u>Introduction</u>. In this report an attempt is made to extend the basic concept of a graph. If we start with a finite nonempty set V, we may pick a set E whose elements are unordered pairs of elements from V and obtain an ordered pair of sets (V,E) such that this yields the usual definition of an ordinary graph, i.e., finite, undirected, and without loops or multiple edges. Since E consists of of two element subsets of elements of V, we raise the question of considering sets $\pi(k)$ whose elements are certain unordered k-tuples of elements from V for $k \geq 2$. By imposing suitable restrictions, $\pi(2)$ turns out to be E. For k = 3 we proceed as follows. If $p \in \pi(3)$, then p = (uvw), where u, v and w are distinct elements from V. The following possibilities now arise for the unordered pairs of elements in p as shown in Figure 1.

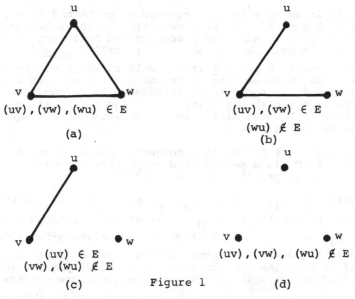

(uv),(vw),(wu) $\in$ E

(a)

(uv),(vw) $\in$ E

(wu) $\notin$ E

(b)

(uv) $\in$ E

(vw),(wu) $\notin$ E

(c)

(uv),(vw), (wu) $\notin$ E

(d)

Figure 1

We shall call elements of type (a) or (b) the elements of the set $\pi(3)$. These elements are called <u>parks</u> and we denote the set $\pi(3)$ as P. Hence we obtain an "extended" graph (V,E,P). Since an ordinary graph can be thought of as being embedded in an extended graph, we will not distinguish between the two and we will use the word graph throughout the report.

2. <u>Basic results and extensions</u>. It is readily seen that parks p_1 and p_2 are identical if and only if they have at least two common edges. This observation leads to the following result.

Proposition 2.1. For any graph G , $|P| = \sum_{v \in V} \binom{\deg v}{2} - 2T$

where T is the number of triangles of G .

An obvious corollary to Proposition 2.1 is that for a bipartite graph, $|P| = \sum_{v \in V} \binom{\deg v}{2}$.

We next define analogues of basic graph theory concepts for extended graphs. Two distinct parks are _adjacent_ if they have a line in common. The _degree_ of a park p in a graph G is the number of distinct parks adjacent to p and will be written as DEG(p) . If for a line x of a graph G , $P(x)$ denotes the number of parks containing the line x , the following result holds.

Proposition 2.2. Let p be any park in a graph G . Then

$$\text{DEG}(p) = \sum_{x \in p} (p(x) - 1) .$$

A _p-walk_ is an alternating sequence of parks and lines, beginning and ending with a park, such that each line in the sequence is contained in the parks immediately preceding and following it in the sequence. A p-walk joining p_1 and p_n is called _closed_ if $p_1 = p_n$, and _open_ otherwise. A _p-trail_ is a p-walk in which no line is repeated. A p-walk in which neither lines nor parks repeat is called a _p-path_. A graph G is _p-connected_ if between every two parks p , p' in G there is a p-path beginning at p and ending at p' . Although it is not true that connectivity and p-connectivity are equivalent, we do have the following result.

Proposition 2.3. (a) If G is a connected graph, then G is p-connected. (b) If G is a p-connected graph each of whose components has at least two lines, then G is connected.

A p-walk $p_1, x_1, p_2, \ldots, p_n, x_n, p_{n+1}$ $(n \geq 3)$ in which no park repeats, and where $p_1 = p_{n+1}$, is called a _p-cycle_ and is denoted $p_1, p_2, \ldots, p_n, p_1$. A graph G is called _p-hamiltonian_ if it has a p-cycle which contains every park of G . A graph G is called a _p-tree_ if it contains no p-cycles. It is readily shown that the only connected graphs with $n \geq 4$ vertices that are p-trees are paths.

The _park connectivity_ $\mu(G)$ of a graph G is the minimum number of parks such that the removal of the lines of these parks will disconnect G . In an attempt to relate park connectivity to connectivity and line connectivity, examples were found to show that $\mu(G)$ and $\varkappa(G)$ are not comparable; however, we do have the following result.

Proposition 2.4. For any graph G ,
    (a) $\mu(G) \leq \lambda(G)$,
    (b) $\mu(G) \leq \left\{ \dfrac{\min \deg G}{2} \right\}$.

We conclude this section by defining the following distance function on the set of parks of a connected graph.

$$
\begin{cases}
0, & \text{if } p_1 = 2 \\
1, & \text{if } p_1 \text{ is adjacent to } p_2 \\
2 + \min\limits_{\substack{u \in p_1 \\ v \in p_2}} \{d(u,v)\}, & \text{otherwise}
\end{cases}
$$

A straight forward argument shows that D is a metric.

3. **Park graphs.** In the literature of graph theory, we come across many graph valued functions defined on the set of all graphs. Among these functions are the line graph, total graph, block graph, block-cut point graph and clique graph functions. The concept of a park in a graph was introduced in the last section, and we now define a graph valued function P (the **park graph function**) on the set of graphs.

The **park graph** of a graph G, denoted $P(G)$, is a graph whose vertex set can be put in 1-1 correspondence with the set of parks of G such that two vertices in $P(G)$ are adjacent if and only if the corresponding parks are adjacent. Some results that follow immediately from this definition are listed in the following proposition.

Proposition 3.1. If G is a graph with park graph $P(G)$, then

$$
(1) \quad |V(P(G))| = \sum_{v \in V(G)} \binom{\deg v}{2} - 2T ,
$$

$$
(2) \quad |E(P(G))| = \sum_{x \in E(G)} \binom{P(x)}{2} ,
$$

where T is the number of triangles of G.

Proof. Part (1) follows from Proposition 2.1. For part (2), let P denote the set of parks in a graph G and let $f: V(P(G)) \to P$ be a 1-1 map that preserves adjacency, and for each line x in $E(G)$, define $S_x = \{u \in V(P(G)) \mid x \in f(u)\}$. Let H_x be the subgraph of $P(G)$ induced by the vertices in S_x. Since every vertex in H_x corresponds to a park in G which contains x, H_x is a complete subgraph of $P(G)$ on $P(x)$ vertices. Since distinct parks can have at most one common line, for $x \neq x'$ in G, H_x and $H_{x'}$ have no common lines. Also, a complete graph on $m \geq 2$ vertices has $\binom{m}{2}$ lines, hence H_x has $\binom{P(x)}{2}$ lines and the total number of lines in $P(G)$ is given by $\sum_{x \in E(G)} \binom{P(x)}{2}$.

We next examine the park graphs of special types of graphs. An application of Proposition 3.1 can be used to prove the following proposition.

Proposition 3.2. Let G be a connected graph on $k \geq 4$ vertices. Then G is a path of length $k - 1$ if and only if $P(G)$ is a path of length $k - 3$.

Proposition 3.3. A graph G is a cycle of length $k \geq 4$ if and only if P(G) is a cycle of length k.

Since Proposition 3.3 gives a class of graphs that are isomorphic to their park graphs, it is natural to ask if there are other graphs that have this property. The answer to this question is affirmative as is illustrated in the graphs of Figure 2.

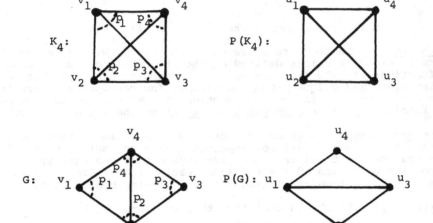

Figure 2

In an attempt to parallel work shown in [1] and [2], it has been possible to show that for a hamiltonian graph G, P(G) is hamiltonian. Although the detailed proof of this is quite lengthy, a sketch of the proof is given below.

We first state a Lemma.

Lemma 3.4. For a graph G, P(G) is hamiltonian if and only if the parks of G can be ordered as $p_1, p_2, \ldots, p_k$, $k \geq 3$, such that p_i is adjacent to p_{i+1}, $i = 1, 2, \ldots, k-1$, and p_k is adjacent to p_1.

Proposition 3.5. If a graph G is hamiltonian, then P(G) is hamiltonian.

To show that P(G) is hamiltonian, we will construct a sequence of parks of G that satisfies the conditions of Lemma 3.4. Since G is hamiltonian, let its vertices be arranged cyclically and be labeled $v_1, v_2, \ldots, v_m$. Label the lines of this cycle $y_1, y_2, \ldots, y_i, \ldots, y_m$ such that y_i joins v_i to v_{i+1}, $i = 1, 2, \ldots, m-1$, and y_m joins v_m to v_1. We will now define the desired sequence in steps.

Step I. The first park in the sequence is $(y_m y_1)$. If $\deg v_1 \leq 3$, the second park in the sequence will be $(y_1 y_2)$, and we proceed to Step II. Otherwise, label the diagonals incident with v_1 and order the parks containing two lines incident with v_1 in the desired manner (Lemma 3.4) such that the park $(y_1 y_2)$ occurs last.

<u>Step II</u>. If deg $v_2 \le 3$, let $(y_2 y_3)$ be the park following $(y_1 y_2)$ and proceed to Step III. Otherwise, label the diagonals incident with v_2 and order the parks containing two lines incident with v_2 in the desired manner such that $(y_2 y_3)$ occurs last.

We now adopt the convention of calling a diagonal that has been labeled a <u>sequential diagonal</u> and a diagonal that has not been labeled a <u>nonsequential diagonal</u>.

<u>Step III</u>. If there are less than two nonsequential diagonals incident with v_3, let the next park in the sequence be $(y_3 y_4)$ and proceed to Step IV. Otherwise, label the nonsequential diagonals incident with v_3 and order the parks containing two nonsequential diagonals (or one nonsequential diagonal and y_2 or y_3) in the desired manner such that $(y_3 y_4)$ occurs last.

<u>Step IV</u>. We now continue the construction in an inductive manner such that the last park in the sequence is adjacent to the first park $(y_m y_1)$.

<u>Step V</u>. If every diagonal is now a sequential diagonal, the proof is complete. Otherwise, we place the parks containing nonsequential diagonals in the sequence already established such that the conditions of Lemma 3.4 are still satisfied. That this can be done is guaranteed by the construction in Steps I - IV. It can be shown that all the parks of the graph G have now been ordered in accordance with Lemma 3.4, hence $P(G)$ is hamiltonian.

4. <u>Planarity</u>. It was shown in the last section that the park graph function preserves certain properties of a graph G while destroying others. In this context, we now consider planarity.

<u>Proposition 4.1</u>. If a graph G is nonplanar, then $P(G)$ is nonplanar.

This can be proven using Kuratowski's criteria for nonplanarity and considering the following cases: (I) G has a subgraph G' homeomorphic from K_5; and (II) G has a subgraph G' homeomorphic from $K(3,3)$. In either case, it can be shown that $P(G)$ has a subgraph homeomorphic from $K(3,3)$ and is therefore nonplanar.

Since the proposition "G planar imples $P(G)$ is planar" is false (the park graph of an n-star, $n > 6$ is nonplanar), it is natural to ask under which conditions this will be true. Although a characterization has not been obtained, it has been shown that for a planar graph G, each of the following conditions is sufficient for $P(G)$ to be planar:
(a) max deg $G \le 2$,

(b) max deg $x \leq 3$,
  $x \in E(G)$
(c) max $P(x) \leq 2$.
  $x \in E(G)$

Furthermore, examples have been found to show that none of the conditions can be relaxed and still imply that $P(G)$ is planar. On the other hand, for a graph G with planar park graph $P(G)$ we must have the following:
(a) max deg $G \leq 4$,
(b) max deg $x \leq 6$,
  $x \in E(G)$
(c) max $P(x) \leq 4$.
  $x \in E(G)$

We now ask if for a graph G with planar park graph $P(G)$ whether or not it might be possible to impose some conditions on the graph G that are more restrictive than those mentioned above. This is answered in the negative in the following proposition. The proof is given by considering counter-examples.

<u>Proposition 4.2.</u> Let G be a graph with park graph $P(G)$. Then $P(G)$ planar fails to imply any of the following:
(a) max deg $G < 4$,
(b) max deg $x < 6$,
  $x \in E(G)$
(c) max $P(x) < 4$.
  $x \in E(G)$

REFERENCES

1. M. Behzad and G. Chartrand, Total graphs and traversability, <u>Proc. Edinburgh Math. Soc.</u>, 15 (1966), 117-120.

2. G. Chartrand, On hamiltonian line-graphs, <u>Trans. Amer. Math. Soc.</u>, to appear.

HAMILTONIAN CIRCUITS IN GRAPHS AND DIGRAPHS

C.St.J.A. Nash-Williams,[1] University of Waterloo

This lecture aims to describe research in which the speaker is currently engaged and at the same time to survey the background to that research. Some results are stated in deliberately vague terms, partly because excessive detail is inappropriate in a single lecture designed to convey the main underlying ideas of the subject, but also partly because the research here described has not yet progressed to the point where exact details could in all cases be supplied. Detailed proofs (and where necessary exact statements) of results believed to be new will in due course be published elsewhere.

I take as my starting-point the following theorem of Dirac [1].

__Theorem 1.__ If a graph G has order $n (\geq 3)$ deg $v \geq \frac{1}{2}n$ for every $v \in V(G)$, then G has a Hamiltonian circuit.

The following is an indication of one method of proof. Let C be one of the longest circuits in G. Suppose that C is not a Hamtonian circuit of G. Then we can select a path A in G such that A is disjoint from C and is, subject to this requirement, as long as possible. If u and v are the end-vertices of A, then the fact that deg $u \geq \frac{1}{2}n$ and deg $v \geq \frac{1}{2}n$ can be used to establish the existence of vertices w_1, w_2 on C which are adjacent to u, v, respectively, and are such that a path of C joining w_1 to w_2 together with A and the edges joining w_1 to u and w_2 to v make up a circuit longer than C. Thus supposing C to be non-Hamiltonian in G leads to a contradiction. (For somewhat different proofs, see [1] and [8].)

One might ask whether the $\frac{1}{2}n$ in the statement of Theorem 1 can be replaced by anything smaller. The answer is that it cannot, since when n is odd there are two ways of giving examples of graphs with n vertices in which all valencies are at least $\frac{1}{2}(n-1)$ but which have no Hamiltonian circuit. One such graph, which we might denote by $K_{(n+1)/2} \vee K_{(n+1)/2}$, is the union of two complete subgraphs which have $\frac{1}{2}(n+1)$ vertices each and just one common vertex. Another, which we might denote by $K^+_{(n+1)/2, (n-1)/2}$, is any graph with $\frac{1}{2}(n+1)$ "red" vertices and $\frac{1}{2}(n-1)$ "blue" vertices in which no two "red" vertices are adjacent, every "red" vertex is adjacent to every "blue" vertex and it is immaterial which pairs of "blue" vertices are

[1]Research supported by grant from the National Research Council of Canada.

adjacent. (According to standard notational conventions, this graph would be denoted by $K_{(n+1)/2, (n-1)/2}$ when no two "blue" vertices are adjacent, and the superscript $+$ is intended to indicate the possible addition to $K_{(n+1)/2, (n-1)/2}$ of some edges joining pairs of "blue" vertices.) However, in a certain sense, these are the only counter-examples to such a strengthening of Dirac's theorem, since one can prove

Theorem 2. If a graph G has n vertices and has no Hamiltonian circuit and $\deg v \geq \frac{1}{2}n - \varepsilon_n$ for every $v \in V(G)$, where ε_n is (in a sense to be made precise) small compared with n, then G is (in a sense to be made precise) very much like either $K_{(n+1)/2} \vee K_{(n+1)/2}$ or some $K^+_{(n+1)/2, (n-1)/2}$.

[It is of course understood that the process of making precise the sense in which G is "very much like" one of these graphs would include some discussion of a mild adjustment of the suffixes so that the assertion makes sense regardless of the parity of n.]

The proof of Theorem 2 consists essentially in thinking out somewhat more fully the implications of the method of proof of Theorem 1 which was sketched above. While ε_n has to be small compared with n, it can nevertheless be made large enough to tend to infinity as n tends to infinity, a fact which is useful in proving the next two theorems.

Let a graph G with n vertices be called a **Dirac graph** if $n \geq 3$ and $\deg v \geq \frac{1}{2}n$ for every $v \in V(G)$. The next two theorems state in effect that, for large graphs, the hypothesis of Dirac's theorem implies somewhat more than the existence of a Hamiltonian circuit.

Theorem 3. If a Dirac graph has more than about 10 vertices, then it has two edge-disjoint Hamiltonian circuits.

The idea of the proof is as follows. Let G be the graph in question: then by Theorem 1 we can find a Hamiltonian circuit C of G. Let $G - E(C)$ denote the graph obtained from G by removing the edges in C. If $G - E(C)$ has a Hamiltonian circuit, then G has two edge-disjoint Hamiltonian circuits and we are done. If $G - E(C)$ has no Hamiltonian circuit, it satisfies the hypotheses of Theorem 2, and thus Theorem 2 gives us some very strong information about the structure of $G - E(C)$. This information is sufficient to enable us to say that, in place of C, we could have selected a more suitable Hamiltonian circuit C' of G which would have ensured that $G - E(C')$ had a further Hamiltonian circuit, as required.

By pursuing the ideas indicated in the preceding paragraph a little further, one can prove

Theorem 4. For every positive integer k, there exists a positive integer n_k such that every Dirac graph with more than n_k vertices has k edge-disjoint Hamiltonian circuits.

The foregoing ideas suggested themselves to me by a somewhat indirect route starting from a consideration of how Theorem 1 can be generalized to digraphs. In discussing digraphs, we shall use the contractions "dipath" and "dicircuit" for "directed path" and "directed circuit". A digraph D is _symmetric_ if, for every pair u, v of distinct vertices of D, either u and v are not joined by any edge at all or they are joined by two edges of which one has tail u and head v while the other has tail v and head u. In many contexts, a symmetric digraph can be regarded as virtually equivalent to a graph, and we embody this idea in the following definition. A _duplicate_ of a graph G is a symmetric digraph D such that (i) V(D) = V(G), (ii) if two vertices are not joined by an edge of G then they are not joined by an edge of D, and (iii) if two vertices u,v are joined by an edge of G then they are joined in D by an edge with tail u and head v and by another edge with tail v and head u. Then obviously D has a Hamiltonian dicircuit if and only if G has a hamiltonian circuit, which shows that Theorem 1 is logically equivalent to the proposition: _if a symmetric digraph_ D _has_ n(≥ 3) _vertices and_ od v ≥ $\frac{1}{2}$n _and_ id v ≥ $\frac{1}{2}$n _for every_ v ∈ V(D) _then_ D _has a Hamiltonian dicircuit_. It follows that the following proposition may be regarded as a generalization of Theorem 1.

Theorem 5. If a digraph D has n vertices and od v ≥ $\frac{1}{2}$n and id v ≥ $\frac{1}{2}$n for every v ∈ V(D), then D has a Hamiltonian dicircuit.

In 1965, at what should now probably be termed the University of Waterloo First Combinatorics Conference, I asked whether Dirac's theorem had been generalized to digraphs in the above sense, and, as no such generalization seemed to be known, I managed after struggling with the problem for about a week to produce a proof of Theorem 5 which has essentially the analogue for digraphs of the proof of Theorem 1 sketched above. This had been submitted for publication, but fortunately not actually published, when I learned that it was not new, since Ghouila-Houri [3] had ingeniously proved the following considerably stronger result.

Theorem 6. If a strongly connected digraph D has n vertices and deg v ≥ n for every v ∈ V(D), then D has a Hamiltonian dicircuit.

It requires only a few moments' thought to see that a digraph with n vertices in which each indegree and each outdegree is greater than or equal to $\frac{1}{2}$n must automatically be strongly connected: thus Theorem 6 is genuinely a generalization of Theorem 5.

As was pointed out by Ghouila-Houri, the following is an immediate consequence of Theorem 6.

Corollary 6a. If a digraph D has n vertices and deg v ≥ n-1 for every v ∈ V(D), then D has a Hamiltonian dipath.

This is proved by considering a digraph D' obtained by adjoining to D a new vertex w and 2n new edges so that for each v ∈ V(D) one of these new edges has tail v and head w and another of them has head v and tail w. A moment's reflection shows that D' satisfies the hypotheses of Theorem 6 (with $n + 1$ replacing n) and so has a Hamiltonian dicircuit, say C. If from C we remove w and the two edges of C incident with w we obtain a Hamiltonian dipath of D.

A special case of Corollary 6a is the following well known theorem of Rédei, elementary proofs of which are given in [4], [7], and [8].

<u>Corollary 6b</u>. Every tournament has a Hamiltonian dipath.

Thus Ghouila-Houri's theorem pleasantly unifies two well known results, Theorem 1 and Corollary 6b.

We remarked that Theorem 1 is equivalent to the proposition that a symmetric digraph with $n(\geq 3)$ vertices in which all indegrees and outdegrees are at least $n/2$ has a Hamiltonian dicircuit. However, if a symmetric digraph has a Hamiltonian dicircuit C, it has two edge-disjoint Hamiltonian dicircuits because, for each edge e in C, the digraph, being symmetric, has another edge e' whose tail and head are respectively the head and tail of e, and the edges e' thus associated with edges e in C determine a second Hamiltonian dicircuit which is edge-disjoint from C. Thus Dirac's theorem is also equivalent to the proposition: if a symmetric digraph D has $n(\geq 3)$ vertices and $od\ v \geq \frac{1}{2}n$ and $id\ v \geq \frac{1}{2}n$ for every $v \in V(D)$ then D has two edge-disjoint Hamiltonian dicircuits. Deleting the word "symmetric" from this proposition would therefore yield another possible generalization of Dirac's theorem which is stronger than Theorem 5, and (as already suggested in the "Unsolved Problems" section of [10]) I think this generalization likely to be true except for a few digraphs with a small number of vertices. In fact, I believe that research undertaken in the last few months indicates promising progress towards a proof of

<u>Conjecture A</u>. If a digraph D has $n(\geq 5)$ vertices and $od\ v \geq \frac{1}{2}n$ $id\ v \geq \frac{1}{2}n$ for every $v \in V(D)$, then D has two edge-disjoint Hamiltonian dicircuits.

The method by which I hope to prove this is similar to that sketched above for Theorem 3, i.e. using Theorem 5 to establish that D has one Hamiltonian dicircuit C and attempting to show that, if $D - E(C)$ has no Hamiltonian dicircuit, then it must be of a special kind, probably bearing a close resemblance to either a duplicate of $K_{(n+1)/2} \vee K_{(n+1)/2}$ or a duplicate of $K_{(n+1)/2,(n-1)/2}$ with possibly additional directed edges joining pairs of "blue" vertices (some mild adjustment of the suffixes being permissible in order to make sense of this statement regardless of the parity of n). However, arguments of this type tend to become substantially more complicated for digraphs, and this proof of Conjecture A, if successfully completed, seems not unlikely to take 100 pages or so! The degree of difficulty of the proposed proof is perhaps only moderate, since it seems to require only a limited number of devices, but these devices have to be applied over and over again to analyze numerous cases. Thus, until better methods can be devised, the task of extending Conjecture A as Theorem 4 extends Theorem 3 would seem to be prohibitive.

It was, however, the investigation of this method of proving Conjecture A which first caused my realization that similar arguments could, with rather less difficulty, be applied to graphs to yield Theorems 3 and 4.

I should like to mention another problem concerning digraphs which seems to be much harder than the corresponding one for graphs. At the age of fourteen, Pósa [9] obtained a nice strengthening of Theorem 1 which roughly speaking says that we can allow the graph to have <u>some</u> vertices of valency less than $\frac{1}{2}n$ provided that there are not too many. A slightly different proof of Pósa's theorem is as follows:

<u>Theorem 7</u>. Let G be a graph with n(≥ 3) vertices such that

   (i) for every positive integer k less than $\frac{1}{2}(n-1)$, the number of vertices of degree less than or equal to k is less than k,

 (ii) the number of vertices of degree less than or equal to $\frac{1}{2}(n-1)$ is less than or equal to $\frac{1}{2}(n-1)$.

Then G has a Hamiltonian circuit.

In the Unsolved Problems section of [2], I conjectured the following analogue for digraphs.

<u>Conjecture B</u>. Let G be a digraph with n vertices such that

   (i) for every positive integer k less than $\frac{1}{2}(n-1)$, the number of vertices of indegree less than or equal to k is less than k,

 (ii) the number of vertices of indegree less than or equal to $\frac{1}{2}(n-1)$ is less than or equal to $\frac{1}{2}(n-1)$,

(iii) for every positive integer k less than $\frac{1}{2}(n-1)$, the number of vertices of outdegree less than or equal to k is less than k,

 (iv) the number of vertices of outdegree less than or equal to $\frac{1}{2}(n-1)$ is less than or equal to $\frac{1}{2}(n-1)$.

Then G has a Hamiltonian dicircuit.

I spent about nine months trying to prove this conjecture, and made very little headway.

The difficulties described in connection with Conjecture A and B suggest that we need improved techniques for handling problems of this type. Personally, I do not feel that I have any strong insight into what these techniques might be. Conceivably, they might involve some method of embedding these "discrete" problems in "continuous" ones and/or some way of relating them to the duality theorem of linear programming.

Two tentative suggestions for future research might be in order. First, it seems, in a sense, disappointing that the inequality in the statement of Theorem 6 is not deg v ≥ n - 1 because then the theorem would also contain another well known theorem on tournaments which asserts that all strongly connected tournaments (with more than one vertex) have Hamiltonian dicircuits ([4], [7]). In fact, Theorem 6 would be false if we changed the inequality as suggested. However, I think there is a very good chance that one could without extreme difficulty show that, with certain specifically identifiable exceptions, all strongly connected digraphs with n(≥ 2) vertices and in which deg v ≥ n - 1 for every vertex v have Hamiltonian dicircuits. This would strengthen Ghouila-Houri's theorem so as to include the theorem on strongly connected tournaments.

Secondly, one way of describing Dirac's theorem would be to say that, if we call a graph (≥ k)-degree when all its vertices have degree greater than or equal to k, the theorem characterizes those (≥ $\frac{1}{2}$n)-degree graphs with n vertices which have Hamiltonian circuits (by showing, in fact, that they <u>all</u> have them if n ≥ 3). Can one improve on this by characterizing, for example, all (≥ $\frac{1}{3}$n)-degree graphs with n vertices which have Hamiltonian circuits? This might be one way to make inroads on the general problem of trying to characterize all graphs with Hamiltonian circuits, a problem which seems to have always intrigued graph-theorists but to be in the present state of knowledge far beyond anyone's capacity.

Incidentally, it might seem at first sight that, even if someone did solve this problem completely, we should still be left with the even more general problem of characterizing digraphs with Hamiltonian dicircuits, since the discussion given earlier in this lecture shows that the problem for graphs can be regarded as a special case of the problem for digraphs. Curiously, however, a solution of the problem for graphs would settle the problem for digraphs also, since for any digraph D one can construct a graph G which has a Hamiltonian circuit if and only if D has a Hamiltonian dicircuit as follows. To each vertex v of D there corresponds a path P_v in G, the paths P_v being disjoint and each of them including three or more vertices. Let the end-vertices of P_v be u_v and w_v. To each directed edge e of D there corresponds an undirected edge x_e of G joining w_{ea} to u_{eb}, where ea, eb denote the tail and head of e respectively; and the paths P_v and edges x_e together make up the whole of G. The reader will easily see that G has the required property. It might finally be noted that if the paths P_v are all taken to be of odd length then G is bipartite, so that, if we determined which bipartite graphs have Hamiltonian circuits, then we should know which digraphs have Hamiltonian dicircuits, which would tell us which symmetric digraphs have Hamiltonian dicircuits, which would tell us which graphs have Hamiltonian circuits. Thus (unlike certain matching problems) the Hamiltonian circuit problem is no easier for bipartite graphs than for graphs in general.

REFERENCES

1. G.A. Dirac, Some theorems on abstract graphs, <u>Proc. London Math.</u>
   Soc. (3) 2 (1952), 69-81.

2. P. Erdös and G. Katona (editors), <u>Theory of Graphs</u>, Proceedings
   of the Symposium at Tihany, Hungary (Publishing House of the
   Hungarian Academy of Sciences, Budapest, and Academic Press, New
   York, 1968).

3. A. Ghouila-Houri, Une condition suffisante d'éxistence d'un circuit hamiltonien, <u>C.R. Acad. Sci. Paris</u> 251 (1960), 495-497.

4. F. Harary, R.Z. Norman and D. Cartwright, <u>Structural Models: An
   Introduction to the Theory of Directed Graphs</u> (John Wiley and
   Sons Inc., New York, 1965).

5. J. Moon and L. Moser, On Hamiltonian bipartite graphs, <u>Israel J.
   Math.</u> 1 (1963), 163-165.

6. C.St.J.A. Nash-Williams, On Hamiltonian circuits in finite graphs,
   <u>Proc. Amer. Math. Soc</u>. 17 (1966), 466-467.

7. O. Ore, <u>Graphs and their uses</u> (Random House Inc., New York, 1963).

8. O. Ore, <u>Theory of Graphs</u>, American Mathematical Society Colloquium Publications Volume XXXVIII (Providence, 1962).

9. L. Pósa, A theorem concerning Hamiltonian lines, <u>Magyar Tud.
   Akad. Mat. Kutató Int. Közl</u>. 7 (1962), 225-226.

10. W.T. Tutte (editor), <u>Recent advances in combinatorics</u>, Proceed-
    ings of a Conference held at Waterloo, Ontario in May 1968 (Aca-
    demic Press, New York) (to appear).

ON THE DENSITY AND CHROMATIC NUMBERS OF GRAPHS

E.A. Nordhaus, Michigan State University

1. <u>Introduction</u>. In this paper, various relationships among four parameters associated with a graph and its complement are examined. These parameters are the density, chromatic number, point independence number and the partition number. Some of the inequalities obtained can be used to determine lower bounds for Ramsey numbers. A relationship is established between the Ramsey numbers and the Zykov numbers. Finally a proof is given that in any cotree the density and chromatic number are equal.

2. <u>Chromatic numbers</u>. For convenience, we set $\chi(G) = k$ and $\chi(\overline{G}) = \overline{k}$ for a graph G. If G has order p, then $1 \le k \le p$ and $1 \le \overline{k} \le p$. Precise bounds for the sum and product of k and $\overline{k}$ were given in [5] by Nordhaus and Gaddum:

$$(1) \qquad \{2\sqrt{p}\} \le k + \overline{k} \le p + 1,$$

$$(2) \qquad p \le k\,\overline{k} \le \left[\left(\frac{p+1}{2}\right)^2\right].$$

The upper and lower bounds occurring in (1) and (2) were shown to be best possible, in the sense that each bound is attained by an infinite number of graphs. It is of interest to note that if $g(x,y)$ and $a(x,y)$ denote respectively the geometric and arithmetic means of positive numbers x and y, that the above inequalities may be concisely written as

$$(3) \qquad g(1,p) \le g(k,\overline{k}) \le a(k,\overline{k}) \le a(1,p).$$

It follows that for an arbitrary graph of order p and chromatic number k, the lattice point $(k,\overline{k})$ lies in a closed region A_p bounded by the straight line $k + \overline{k} = p + 1$ and the hyperbola $k\overline{k} = p$. Conversely, as pointed out by Stewart [7], corresponding to every lattice point $(k,\overline{k})$ in A_p there exists at least one graph G of order p for which $\chi(G) = k$ and $\chi(\overline{G}) = \overline{k}$, and such a graph is constructed. This observation was also made by Finck [2], who investigated the properties of graphs having lattice points $(k,\overline{k})$ on or near the boundary of the region A_p.

If $L(p)$ denotes the number of lattice points $(k,\overline{k})$ in the closed region A_p, then

$$(4) \qquad L(p) = \binom{p+1}{2} - \sum_{k=1}^{p-1} \tau(k)$$

where $\tau(k)$ is the number of distinct positive divisors of k. To establish (4), we note that for a fixed integer k, $2 \le k \le p-1$, the number of lattice points $\lambda(k)$ in A_p which are not on the hyperbola $k\overline{k} = p$ is $\lambda(k) = p+1-k-[p/k]$. Since $\lambda(1) = \lambda(p) = 0$, then

$$L(p) = \tau(p) + \sum_{k=1}^{p} \lambda(k) = \tau(p) + \binom{p+1}{2} - \sum_{k=1}^{p} [p/k].$$

Using the known number-theoretic relation $\sum_{k=1}^{p} \tau(k) = \sum_{k=1}^{p} [p/k]$, one obtains (4). The number $L(p)$ affords a lower bound for the number of non-isomorphic graphs of order p, but is not very satisfactory since in general many non-isomorphic graphs have the same parameters p, k, and $\overline{k}$.

3. <u>Density</u>. The concept of density (or clique number) of a graph was introduced by Zykov [8]. If a graph G of order p contains a complete subgraph of order ρ but no complete subgraph of order $\rho + 1$, then $\rho = \rho(G)$ is called the <u>density</u> of G. Clearly the density ρ satisfies the inequalities $1 \le \rho \le k \le p$, since at least ρ colors are needed to color G. Zykov has shown that there is no further relation possible between ρ and k by constructing for each graph G a graph H of greater order than that of G for which $\rho(H) = \rho(G)$ and $\chi(H) = \chi(G) + 1$. By iteration of this construction, one obtains a graph for which $k - \rho$ is arbitrarily large.

The graphs constructed by Stewart [7] corresponding to an arbitrary lattice point $(k,\overline{k})$ in A_p have the additional property that $\rho = k$ and $\overline{\rho} = \overline{k}$. It follows for an arbitrary graph G of order p having density ρ and chromatic number k that

(5) $$\rho + \overline{\rho} \le k + \overline{k} \le p + 1, \quad \rho\overline{\rho} \le k\overline{k} \le \left[\left(\frac{p+1}{2}\right)\right]^2,$$

and the upper bounds given in equations (1) and (2) are sharp for the sum and product of the densities ρ and $\overline{\rho}$. However the lower bounds are no longer best possible relative to density, as shown by taking $G = C_5$, a cycle on five points. Then $\rho = \overline{\rho} = 2$, and the lattice point $(2,2)$ fails to lie in the region A_5. Certain classes of graphs for which $\rho = k$ or $\overline{\rho} = \overline{k}$ or where both equalities hold have been considered by Berge [1].

4. <u>The point independence number</u>. A set of points of a graph G is called <u>independent</u> if no two are adjacent. The maximum number of points in any independent set is called the <u>point independence number</u> and is denoted by $\beta_0 = \beta_0(G)$. It is immediately apparent that $\rho(G) = \beta_0(\overline{G})$ and $\rho(\overline{G}) = \beta_0(G)$, that is, the density of any graph is equal to the point independence number of the complementary graph. Furthermore, by a result due to Ore [6, p. 225], $p \le k\beta_0$. Similarly $p \le \overline{k}\overline{\beta}_0$, or in terms of density, $p \le k\overline{\rho}$ and $p \le \overline{k}\rho$. Then $\{p/k\} \le \rho \le k$ and $\{p/\overline{k}\} \le \overline{\rho} \le \overline{k}$. These relations provide lower bounds for $\rho + \overline{\rho}$ and $\rho\overline{\rho}$ which depend on k and $\overline{k}$ as well as p:

(6) $$\{p/k\} \cdot \{p/\overline{k}\} \le \rho\overline{\rho}, \quad \{p/k\} + \{p/\overline{k}\} \le \rho + \overline{\rho}.$$

The inequalities (5) and (6) constitute for density analogs to the equations (1) and (2) found for chromatic numbers.

5. <u>The partition number</u>. The <u>partition number</u> $\pi = \pi(G)$ is the minimum number of point disjoint complete graphs which cover the points of G. Such a cover will be called a π-cover. We next prove that $\pi = \bar{k}$ and $\bar{\pi} = k$, so the partition number of a graph equals the chromatic number of the complementary graph. Consider a k-coloring of the points of a graph G having chromatic number k. The k color classes are sets of independent points, so in $\bar{G}$ there arise k disjoint complete graphs covering the points of $\bar{G}$, and $\bar{\pi} \leq k$, by the minimum property of the partition number. Next consider a decomposition of the vertex set of G into a minimum number π of disjoint complete graphs. Each of these π sets of points is an independent set in $\bar{G}$. If we use these sets as color classes, we have a π-coloring of $\bar{G}$ and $\bar{k} \leq \pi$ by the minimum property of the chromatic number. Hence $\bar{\pi} = k$, and similarly $\pi = \bar{k}$.

The above results are primarily useful for determining the parameters of $\bar{G}$ when those of G are known, or vice-versa. Thus $\bar{\rho} = \beta_o$, $\bar{k} = \pi$, $\bar{\beta}_o = \rho$, and $\bar{\pi} = k$.

6. <u>Ramsey and Zykov numbers</u>. The determination of the density of a graph or its complement is closely related to the difficult combinatorial problem of determining Ramsey numbers. The <u>Ramsey number</u> $R(m,n)$ is usually defined as the smallest positive integer p such that every graph G of order p or greater contains a complete graph K_m or a set of n independent points. In terms of density, every graph G of order p with $p \geq R(m,n)$ has $\rho(G) \geq m$ or $\rho(\bar{G}) \geq n$. It is easy to show, for example, that $R(3,3) = 6$. Additional properties of the Ramsey numbers can be found in the paper [3] by Greenwood and Gleason. Only a few non-trivial values of the Ramsey number are known. (see Kalbfleish [4]).

Zykov [8] has defined numbers $Z(m,n)$ related to the Ramsey numbers as follows: $Z(m,n)$ is the maximum density of the sum (union) of two graphs of densities m and n; i.e. the maximum order of a complete graph which can be obtained by the addition of two graphs of densities m and n. We assume that both summands have the same points, $Z(m,n)$ in number, since the addition of isolated points to either summand does not affect its density. For example, $Z(2,2) = 5$, since the decomposition of K_5 into two cycles of length 5 shows that $Z(2,2) \geq 5$ and the fact that every graph of order 6 has $\rho \geq 3$ or $\rho \geq 3$ shows $Z(2,2) < 6$.

An obvious conjecture is that $Z(m,n) + 1 = R(m+1,n+1)$, and this can be shown to be correct by establishing the inequalities $Z(m,n) \leq R(m+1,n+1) - 1$ and $R(m+1,n+1) - 1 \leq Z(m,n)$. These inequalities follow at once from the definitions.

Consider a collection of point disjoint graphs G_i $(i = 1,2,\ldots, n)$ having parameters p_i, ρ_i, k_i, β_{oi}, π_i and let G^i be the join of the G_i, with parameters p, ρ, k, β_o, and π. It readily follows that these parameters have the values $p = \Sigma p_i$, $\rho = \Sigma \rho_i = \bar{\beta}_o$, $k = \Sigma k_i = \bar{\pi}$, $\beta_o = \max \beta_{oi} = \bar{\rho} = \max \bar{\rho}_i$, $\pi = \bar{k} = \max \bar{k}_i = \max \pi_i$, and that $Z(\rho,\bar{\rho}) \geq p$. Good lower bounds for $Z(\rho,\bar{\rho})$ depend on judicious choices for the graph G_i. In particular, they should be chosen so that for the graph G the inequality $\overline{\rho}\rho < p$ will hold, since in general $Z(\rho,\rho) \geq \rho\rho$.

7. <u>Trees, forests and cotrees</u>. In section 3 the existence of a class of graphs having equal density and chromatic number was noted. This class clearly includes the complete graphs and bipartite graphs. It also includes graphs for which $k\overline{k} = p$, since this relation and the inequalities $p \leq k\rho$ and $p \leq \overline{k}\overline{\rho}$ of section 5 imply $k \leq \rho$ and $\overline{k} \leq \overline{\rho}$, so that $k = \rho$ and $\overline{k} = \overline{\rho}$.

The complement of a connected acyclic graph (a tree) is called a <u>cotree</u>, and the remainder of this section is devoted to showing that every cotree has equal density and chromatic number. The well known theorem of Gallai which states that in any connected graph G of order p the equations $\alpha_0 + \beta_0 = p = \alpha_1 + \beta_1$ must hold, will be useful. Here α_0 and α_1 are respectively the point and line covering numbers, that is, the minimum number of points (lines) in any point (line) cover of G. A point and an incident line are said to cover each other. The parameter β_0 is the point independence number of lines in any independent set of lines of G (no two adjacent).

If G is a connected graph with density 2, any covering of the points of G by disjoint complete graphs employs only complete graphs of types K_1 or K_2. Since each complete graph in such a covering contains at most one point of any independent set of points of G, then $\beta_0 \leq \pi$. This inequality is of course equivalent to $\overline{\rho} \leq \overline{k}$. Next consider a maximum set of β_1 independent lines of G. A covering of the points of G by disjoint complete graphs is obtained by covering the $2\beta_1$ end points of the β_1 independent lines by β_1 graphs of type K_2, and each of the remaining $p - 2\beta_1$ points by a graph of type K_1. Then $\pi \leq \beta_1 + (p - 2\beta_1) = p - \beta_1 = \alpha_1$ by the minimal property of π and by Gallai's theorem. Thus, for any tree, $\beta_0 \leq \pi \leq \alpha_1$. We will show that in any tree, $\beta_1 = \alpha_0$, so by Gallai's theorem $\beta_0 = \alpha_1$. This implies that $\beta_0 = \pi = \alpha_1$, or $\overline{\rho} = \overline{k} = \alpha_1$, proving that the density and chromatic number of any cotree are equal.

It remains to show for any tree T that $\beta_1(T) = \alpha_0(T)$. We prove a more general result: for any forest F, a graph whose components are trees, $\beta_1(F) = \alpha_0(F)$. We use induction on the order p of the forest F. If $p = 1$, then F is a trivial tree and we define $\beta_1 = \alpha_0 = 0$. When $p = 2$, F is a tree with two points or a forest consisting of two trivial trees, and $\beta_1 = \alpha_0$ in each case. We next assume that for all forests with p points that $\beta_1 = \alpha_0$, so in particular this is true for all trees of order p. Consider a forest F of order $p + 1$, where $p \geq 2$. If F has more than one component, then each component is a tree of order p or less, and since $\beta_1 = \alpha_0$ for each component, then $\beta_1(F) = \alpha_0(F)$, since these numbers are found by summing β_1 and α_0 over all components of F. If F is a tree of order $p + 1$, let uv be a line of F, where v is any endpoint of the tree. Remove point u and all lines incident to u. The resulting graph G is a forest of order p containing at least one isolated point, namely point v, and $\beta_1(G) = \alpha_0(G)$ by the inductive hypothesis. If we replace point u and the deleted lines, then $\alpha_0(F) = \alpha_0(G) + 1$ since point u covers all lines incident with point u, and $\beta_1(F) = \beta_1(G) + 1$, since line uv can be added to any set of $\beta_1(G)$ independent lines of G, and no further lines can be added. Thus $\beta_1(F) = \alpha_0(F)$ and the induction is complete.

REFERENCES

1. C. Berge, Some classes of perfect graphs, <u>Graph Theory and Theoretical Physics</u> (edited by F. Harary) Academic Press (1967), 155-165.

2. H.J. Finck, On the chromatic numbers of a graph and its complement, <u>Theory of Graphs</u> (edited by P. Erdös and G. Katona) Academic Press (1968), 99-113.

3. R.E. Greenwood and A.M. Gleason, Combinatorial relations and chromatic graphs, <u>Canad. J. Math.</u> 7, no. 1 (1955), 1-7.

4. J.B. Kalbfleisch, Upper bounds for some Ramsey numbers, <u>Journal of Combinatorial Theory</u>, 2(1967), 35-42.

5. E.A. Nordhaus and J.W. Gaddum, On complementary graphs, <u>Amer. Math. Monthly</u> 63 (1956), 176-177.

6. O. Ore, <u>Theory of Graphs</u>, Amer. Math. Society Colloq. Publications, vol. 38, 1962.

7. B.M. Stewart, On a theorem of Nordhaus and Gaddum, <u>J. Combinatorial Theory</u>, 6 (1969), 217-218.

8. A.A. Zykov, On some properties of linear complexes, <u>Math. Sbornik</u> 24 (1949), 163-188. [Amer. Math. Society Translations No. 79, 1952].

METHODS FOR THE ENUMERATION OF MULTIGRAPHS[1]

Edgar M. Palmer, Michigan State University

One of the most important concepts in graphical enumeration as well as combinatorial analysis is that of a symmetric function called the "cycle index of a permutation group". Although the use of these functions is implicit in the work of earlier authors, the credit for their discovery evidently belongs to J.H. Redfield (1927), who called them "group reduction functions" or just G.R.F.'s. Later they were rediscovered independently by G. Pólya (1937), who used them to obtain a variety of interesting combinatorial results.

In order to illustrate the fundamental role of the cycle index in solving combinatorial problems, we shall examine some of the methods for determining the number of multigraphs having a prescribed number of points and lines. The discussion of several methods serves to emphasize the diverse ways in which cycle indexes can be interpreted and expressed in order to obtain results.

1. <u>Cycle indexes and permutation groups</u>. Let A be a permutation group of order $|A|$ with object set $X = \{1, 2, \ldots, m\}$. The <u>degree</u> of A is the number m of elements in X. For each permutation α in A, let $j_k(\alpha)$ be the number of cycles of length k in the disjoint cycle decomposition of α. The <u>cycle index of A</u>, denoted $Z(A)$, is the polynomial in the variables $a_1, a_2, a_3, \ldots$ given by

(1)
$$Z(A) = \frac{1}{|A|} \sum_{\alpha \in A} \prod_{k=1}^{m} a_k^{j_k(\alpha)} .$$

In order to display the variables, we often write

(2)
$$Z(A) = Z(A; a_1, a_2, \ldots, a_m) .$$

If $f(x)$ is a power series in x, $Z(A, f(x))$ is the series obtained by replacing each variable a_k in $Z(A)$ by $f(x^k)$. That is

(3)
$$Z(A, f(x)) = Z(A; f(x), f(x^2), \ldots, f(x^m)) .$$

Now let B be another permutation group with object set $Y = \{1, 2, \ldots, n\}$. If A and B are isomorphic as abstract groups, we shall write $A \cong B$. Suppose there is an isomorphism from A onto B denoted by a prime, i.e. for each permutation α in A, the image of α is α'. If there is also a 1-1 correspondence $\varphi: X \to Y$ such that for each α in A and each x in X

---

[1]Work supported in part by a graph from the National Science Foundation.

(4) $$\varphi(\alpha x) = \alpha'\varphi(x) \ ,$$

then A and B are said to be <u>identical</u> (or isomorphic as permutation groups) and we write A = B.

Note that if A $\cong$ B, it does not necessarily follow that Z(A) =
Z(B). Furthermore if two groups have the same cycle index, they need
not be identical (see [7, p. 446]).

We shall be concerned especially with the cycle indices of cyclic groups. Let α be a permutation of m objects which has order
r and let Cyc(α) denote the cycle index of the cyclic group generated by α. Then it is easily seen that

(5) $$Cyc(\alpha) = \frac{1}{r} \sum_{i=1}^{r} \prod_{k=1}^{r} a_{a_k}^{(k,i)j_k(\alpha)}/(k,i) \ ,$$

where (k,i) is the g.c.d. of k and i.

If α consists only of a cycle of length r, then Redfield's
formula for Cyc(α) is also easily verified:

(6) $$Cyc(\alpha) = \frac{1}{r} \sum_{d \mid r} \varphi(d) \ a_d^{r/d} \ ,$$

where φ is the Euler φ-function.

2. <u>Multigraphs</u>. Let $V = \{v_1, v_2, \ldots, v_n\}$ be a set of n objects and let $V^{(2)}$ be the collection of all 2-subsets of V. A
<u>multigraph</u> G is a function from $V^{(2)}$ into the non-negative inte-
gers. The elements of V are called the <u>points</u> of G. Any two
points v_i and v_j are called <u>adjacent</u> if and only if $G(\{v_i,v_j\})$ >
0; and $G(\{v_i,v_j\})$ is often called "the number of lines joining v_i
and v_j". Thus the number of lines in G is just $\sum G(S)$, where the
sum is over all elements S of $V^{(2)}$.

Two multigraphs G and H, each with V as the set of points,
are <u>isomorphic</u> if there is a permutation α of V such that for all
i $\neq$ j

(7) $$G(\{v_i,v_j\}) = H(\{\alpha v_i, \alpha v_j\}) \ .$$

Let m and n be integers with m $\geq$ 0 and n $\geq$ 2. We define
g(n,m) to be the number of isomorphism classes of multigraphs having
exactly n points and m lines.

It is easily seen, for example, that g(4,4) = 11 and all 11
multigraphs are shown in Figure 1.

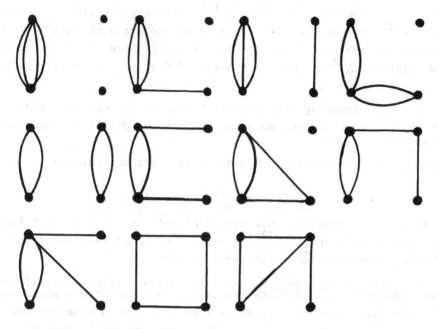

Figure 1

One approach to computing the numbers $g(n,m)$ of multigraphs is to find for each integer $n \geq 2$ an expression for the power series in x which has $g(n,k)$ as the coefficient of x^k for each integer $k \geq 0$. For this purpose we let

$$(8) \qquad g_n(x) = \sum_{k=0}^{\infty} g(n,k) x^k$$

for $n \geq 2$, and we let $g_1(x) = 1$.

Next we consider a unary operation on permutation groups which will enable us to obtain a formula for $g_n(x)$. Let A be a permutation group with object set $X = \{1,2,\ldots,n\}$. The pair group of A, denoted by $A^{(2)}$, is that permutation group whose object set is $X^{(2)}$, the 2-subsets of X, and whose permutations are induced by those in A. That is, for each permutation α in A, there is a permutation α' in $A^{(2)}$ such that for each element $\{i,j\}$ of $X^{(2)}$,

$$(9) \qquad \alpha'\{i,j\} = \{\alpha i, \alpha j\} .$$

If $n = 1$, then $X^{(2)}$ is empty, but for convenience we let

$Z(A^{(2)}) = 1$. When $n = 2$, then $X^{(2)}$ has only one element and so we let $A^{(2)}$ be the identity group on one object with $Z(A^{(2)}) = a_1$. It is easily seen that when $n = 3$, we always have $A = A^{(2)}$ and hence $Z(A) = Z(A^{(2)})$. On the other hand for $n \geq 4$, we always have $A \cong A^{(2)}$ but never is $A = A^{(2)}$.

With S_n denoting the symmetric group on n objects, the forula below for $g(x)$ follows upon the application of Pólya's enumeration theorem (see [2, 6]).

Theorem 1. The generating function $g_n(x)$ for multigraphs is given by

(10) $$g_n(x) = Z(S_n^{(2)}, \; 1/(1-x)) \; .$$

In order to complete this method of solution we require a formula for the cycle index $Z(S_n^{(2)})$. Various means of calculating this cycle index are provided in the next section.

3. **Cycle index formulas.** We shall discuss only three different methods for obtaining formulas for the cycle index of the pair group $S_n^{(2)}$. The first was found by Pólya (see [6]) and is probably the easiest to apply. The partitions of n are denoted by vectors

$(j_1, j_2, \ldots, j_n)$ where $\displaystyle\sum_{k=1}^{n} k j_k = n$, so that j_k is the number of

parts equal to k. We use (r,s) and $[r,s]$ to denote the g.c.d. and the l.c.m. respectively of r and s. Then Pólya's formula for $Z(S_n^{(2)})$ takes the following form:

(11) $$Z(S_n^{(2)}) = \sum_{(j)} \frac{1}{\prod k^{j_k} \cdot j_k!} \prod_{k \text{ odd}} (a_k^{(k-1)/2})^{j_k}$$

$$\prod_{k \text{ even}} (a_k^{(k-2)/2} a_{k/2})^{j_k} \prod_{s < t} a_{[s,t]}^{(s,t) j_s j_t} \prod_k a_k^{k\binom{j_k}{2}} \; ,$$

where the sum is over all partitions $(j) = (j_1, j_2, \ldots, j_n)$ of n.

For example, the partitions of 4 are given by the vectors $(4,0,0,0)$, $(2,1,0,0)$, $(1,0,1,0)$, $(0,2,0,0)$, and $(0,0,0,1)$. From formula (11) we obtain

(12) $$Z(S_4^{(2)}) = \frac{1}{6!} a_1^6 + \frac{3}{8} a_1^2 a_2^2 + \frac{1}{3} a_3^2 + \frac{1}{4} a_2 a_4 \; .$$

From Theorem 1 we have

$$g_4(x) = Z(S_4^{(2)}, \; 1/1-x) \; ,$$

and upon carrying out the substitution of $1/(1-x)$ in the required manner, we find

(13)
$$g(x) = 1+x+3x^2+6x^3+11x^4+18x^5+32x^6+48x^7+75x^8$$
$$+ 111x^9+160x^{10}+\ldots \; .$$

This completes the discussion of the first method for obtaining the coefficients of $g_n(x)$. Further details and a proof of Theorem 1 may be found in [2].

J.H. Redfield considered the pair group $S_n^{(2)}$ from an entirely different point of view. Given two permutation groups A and B, each having the same object set, he constructed a "derived group" and provided a method for obtaining its cycle index from the cycle indexes $Z(A)$ and $Z(B)$. For suitable A and B, the derived group is identical to $S_n^{(2)}$.

Let A and B be permutation groups with object set $X = \{1,2,\ldots,n\}$. The symmetric group S_n also has object set X. The derived group of A and B is denoted by A/B and has as its object set the right cosets of S_n modulo B. For each permutation α in A, there is a permutation α' in A/B such that for any right coset γB of S_n, the image of γB under α' is $\alpha\gamma B$. That is

(14)
$$\alpha'(\gamma B) = \alpha\gamma B.$$

Thus the permutations in A/B consist of all those permutations of the cosets which are induced by A under left multiplication. Hence A/B is a homomorphic image of A.

The number of cosets in the object set of A/B is just the index of B in S_n. Therefore the degree of A/B is $n!/|B|$.

In order to express the pair group $S_n^{(2)}$ as a derived group, we require another binary operation on permutation groups. We let A and B be permutation groups with object sets X and Y respectively. The sum of A and B, denoted $A + B$, is a permutation group whose object set is the disjoint union $X \cup Y$. The permutations in $A + B$ consist of the ordered pairs, written $\alpha\beta$, of permutations α in A and β in B. For each element z of $X \cup Y$, the image, $\alpha\beta(z)$, of z under $\alpha\beta$ is given by

(15)
$$\alpha\beta(z) = \begin{cases} \alpha z & \text{if } z \in X \\ \beta z & \text{if } z \in Y \end{cases} .$$

Thus the order of $A + B$ is $|A||B|$ and the degree is $|X| + |Y|$. Furthermore it is easy to see that

(16)
$$Z(A+B) = Z(A)Z(B) .$$

As before we let S_n, the symmetric group on $n \geq 3$ objects, act on $X = \{1,2,\ldots,n\}$. With $X_1 = \{1,2\}$ and $X_2 = \{2,3,\ldots,n\}$, we have $X_1 \cap X_2 = \emptyset$ and $X_1 \cup X_2 = X$. Next we let the symmetric groups S_2 and S_{n-2} have object sets X_1 and X_2. Then the sum $S_2 + S_{n-2}$ has object set X, and we have the following important identity:

$$(17) \qquad S_n^{(2)} = S_n/(S_2+S_{n-2}) \ .$$

Note that the degree of $S_n/(S_2+S_{n-2})$ is $\binom{n}{2}$. A 1-1 map φ from the $\binom{n}{2}$ cosets modulo $S_2 + S_{n-2}$ to $X^{(2)}$ is defined as follows. For each element γ of S_n,

$$(18) \qquad \varphi(\gamma(S_2+S_{n-2})) = \{\gamma(1),\gamma(2)\}.$$

The identity (17) is then verified by showing that φ and the obvious group isomorphism satisfy equation (4).

One method, though an impractical one, to obtain $Z(A/B)$ from $Z(A)$ and $Z(B)$ makes use of the classical formula for the character of an induced representation. This formula is given in the following theorem, which is essentially Theorem 16.7.2 of [1].

Theorem 2. For each permutation α in A, let C_α be the set of conjugates of α in A, and let α' be the corresponding permutation in A/B. Then the number, $j_1(\alpha')$, of cosets fixed by α' is

$$(19) \qquad j_1(\alpha') = |A| \, |C_\alpha \cap B|/(|B| \, |C_\alpha|).$$

Formula (19) is of some value when $A = S_n$. In this case, suppose α is an element of S_n and for each k, let $j_k = j_k(\alpha)$, the number of k-cycles in α. Then we have

$$(20) \qquad |A|/|C_\alpha| = \prod_{k=1}^{n} k^{j_k} j_k! \ ,$$

and $|C_\alpha \cap B|/|B|$ is just the coefficient in $Z(B)$ of $\prod_{k=1}^{n} a_k^{j_k}$. But it remains to find $j_k(\alpha')$ when $k > 1$. This can be accomplished by observing that

$$(21) \qquad j_1((\alpha')^k) = \sum_{s|k} s j_s(\alpha').$$

Hence, using möbius inversion, $j_k(\alpha')$ can be expressed in terms of the fixed elements of certain powers of α':

$$(22) \qquad j_k(\alpha') = \frac{1}{k} \sum_{s|k} \mu(k/s) j_1((\alpha')^s).$$

Even in the case under consideration when $A = S_n$ and

$B = S_2 + S_{n-2}$, the application of formulas (19) and (22) is a tedious task. A more flexible approach toward finding $Z(A/B)$ which is suitable for any A and B, and which is much easier to apply, was formulated by Redfield.

In order to describe Redfield's method for calculating $Z(A/B)$, we first define the "cup" operation $\cup$ introduced in [7]. Let R be the ring of rational polynomials in the variables $a_1, a_2, a_3, \ldots$. For any two monomials $a_1^{j_1} a_2^{j_2} \ldots a_m^{j_m}$ and $a_1^{i_1} a_2^{i_2} \ldots a_n^{i_n}$ in R we let

$$(23) \qquad a_1^{j_1} a_2^{j_2} \ldots a_m^{j_m} \cup a_1^{i_1} a_2^{i_n} = (\Pi_k k^{j_k} \cdot j_k!) a_1^{j_1} a_2^{j_2} \ldots a_m^{j_m}$$

if the two monomials are identical and it is zero otherwise. The operation is now extended linearly to $R \otimes R$. Since $\cup$ is associative, as well as commutative, it can be extended further to the product $R \otimes \ldots \otimes R$ of any length. This operation was designed for use in conjunction with cycle indexes and, as such, can be interpreted as an inner product of group characters. Redfield exploited it to obtain numerous interesting combinatorial results including a "decomposition theorem", which allows $\cup$-products of cycle indexes to be written as sums of cycle indices [7, p. 445; 3, p. 381]. The following result is a consequence of this decomposition theorem and appears in [7, p. 449] in quite different form.

Theorem 3. Let B be a permutation group of degree n and let α be a permutation of n symbols which has order r. Then

$$(24) \qquad Z(B) \cup Cyc(\alpha) = \sum_{k|r} i_k\, Cyc(\alpha^k) ,$$

where the i_k are uniquely determined non-negative integers.

It can be seen from the decomposition theorem that $Z(B) \cup Cyc(\alpha)$ must be a sum of cycle indexes of groups which are subgroups of both B and the cyclic group generated by α. The coefficients i_k are unique because the cycle indexes $Cyc(\alpha^k)$ are independent. Formula (24) assumes great importance for our purposes because it is easy, as we shall demonstrate, to find the coefficients i_k. Furthermore these coefficients completely determine the cycle structure of α' in A/B. This observation is not hard to establish and is summarized as follows.

Corollary 1. The permutation α' in A/B has i_k cycles of length k for each $k|r$, where the i_k are the coefficients of the $Cyc(\alpha^k)$ in the decomposition of $Z(B) \cup Cyc(\alpha)$.

Thus the contribution to $Z(A/B)$ of α' is $\Pi_{k|r} a_k^{i_k}$. To illustrate, we now proceed to find $Z(S_6/(S_2+S_4))$ using Theorem 3 and its corollary.

From the well known formula for the cycle index of the symmetric groups [7] and from formula (16) for the cycle index of the sum we have

(25)
$$Z(S_6) = \tfrac{1}{6!} \, (a_1^6 + 120a_1a_2a_3 + 40a_1^3a_3 + 15a_1^4a_2 + 120a_6$$

$$+ 40a_3^2 + 15a_2^3 + 144a_1a_5 + 90a_2a_4 + 45a_1^2a_2^2 + 90a_1^2a_4)$$

(26)
$$Z(S_2)Z(S_4) = \tfrac{1}{48} \, (a_1^6 + 7a_1^4a_2 + 8a_1^3a_3 + 9a_1^2a_2^2 + 6a_1^2a_4$$

$$+ 8a_1a_2a_3 + 3a_2^3 + 6a_2a_4) \ .$$

Now suppose α is a permutation in S_6 with cycle structure $a_1a_2a_3$; i.e., $j_1(\alpha) = j_1(\alpha) = j_3(\alpha) = 1$. Then the order of α is 6 and the divisors of 6 are 1,2,3, and 6. Therefore from Theorem 3 we have

(27) $Z(S_2)Z(S_4) \, \cup \, \mathrm{Cyc}(\alpha) = i_1\mathrm{Cyc}(\alpha) + i_2\mathrm{Cyc}(\alpha^2) + i_3\mathrm{Cyc}(\alpha^3) + i_6\mathrm{Cyc}(\alpha^6)$.

On the other hand, since we have from (5)

(28) $\mathrm{Cyc}(\alpha) = \tfrac{1}{6} \, (a_1^6 + 2a_1a_2a_3 + 2a_1^3a_3 + a_1^4a_2)$,

we find that

(29) $Z(S_2)Z(S_4) \, \cup \, \mathrm{Cyc}(\alpha) = \tfrac{5}{2} \, a_1^6 + \tfrac{1}{3} \, a_1a_2a_3 + a_1^3a_3 + \tfrac{7}{6} \, a_1^4a_2$.

Combining (27) and (29) yields

(30) $\displaystyle\sum_{k|6} i_k\mathrm{Cyc}(\alpha^k) = \tfrac{5}{2} \, a_1^6 + \tfrac{1}{3} \, a_1a_2a_3 + a_1^3a_3 + \tfrac{7}{6} \, a_1^4a_2$.

The term $a_1a_2a_3$ appears in the left side of (30) only in $i_1\mathrm{Cyc}(\alpha)$. Hence the coefficient of $a_1a_2a_3$ in the left side of (30) is $i_1/3$. Since its coefficient in the right side of (30) is $1/3$, we have $i_1 = 1$. Subtracting $\mathrm{Cyc}(\alpha)$ from both sides of (30) gives

(31) $i_2\mathrm{Cyc}(\alpha^2) + i_3\mathrm{Cyc}(\alpha^3) + i_6\mathrm{Cyc}(\alpha^6) = \tfrac{14}{6} \, a_1^6 + \tfrac{2}{3} \, a_1^3a_3 + a_1^4a_2$.

The term $a_1^3a_3$ appears in the left side of (31) only in $i_2\mathrm{Cyc}(\alpha^2) = \tfrac{i_2}{3} \, (a_1^6 + 2a_1^3a_3)$. Hence, equating coefficients again we have $2i_2/3 = 2/3$ and so $i_2 = 1$. Subtracting $\mathrm{Cyc}(\alpha^2)$ from both sides of (31) gives

(32) $i_3\mathrm{Cyc}(\alpha^3) + i_6\mathrm{Cyc}(\alpha^6) = 2a_1^6 + a_1^4a_2$.

Equating coefficients of $a_1^4a_2$, we have $i_3/2 = 1$ and so

$i_3 = 2$. Subtracting $2 \operatorname{Cyc}(\alpha^3)$ from both sides of (32) gives

(33)
$$i_6 \operatorname{Cyc}(\alpha^6) = a_1^6 .$$

Therefore $i_6 = 1$ and the cycle structure of α' in $S_6/(S_2+S_4)$ is given by $a_1 a_2^2 a_3 a_6$. Furthermore, since α^2 has structure $a_1^3 a_3$, $(\alpha^2)' = (\alpha')^2$ has structure $a_1^3 a_3^4$, and since α^3 has structure $a_1^4 a_2$, $(\alpha^3)' = (\alpha')^3$ has structure $a_1^7 a_2^4$. Obviously $(\alpha^6)'$ has structure a_1^{15}. At this point we have completely determined the cycle structure of those permutations in $S_6/(S_2+S_4)$ which are induced by the permutations in S_6 which have cycle structure $a_1^6, a_1 a_2 a_3, a_1^3 a_3$ and $a_1^4 a_2$. We can continue in this manner to determine the structure of the others by selecting elements of S_n which generate maximal cyclic subgroups and applying the same process to them.

On completing this process we find

(34)
$$Z(S_6/(S_2+S_4)) = \frac{1}{6!} (a_1^{15} + 120a_1 a_2 a_3^2 a_6 + 40a_1^3 a_3^4 + 15a_1^7 a_2^4$$

$$+ 120a_3 a_6^2 + 40a_3^5 + 60a_1^3 a_2^6 + 144a_5^3 + 180a_1 a_2 a_4^3) .$$

Redfield used this method to find $Z(S_n/(S_2+S_n))$ for $n = 1$ through 7 (see [7, pp. 451-453]).

4. **Multigraphs as unions of graphs.** We have defined multigraphs as functions, and indeed this point of view is helpful in applying Pólya's theorem to obtain Theorem 1. A multigraph with $n \geq 2$ points and m lines can also be considered, however, as a union of m graphs, each with n points and exactly one line. Note that the cycle index of the automorphism group of each of these m graphs is $Z(S_2)Z(S_{n-2})$. By considering multigraphs as unions of graphs one can apply a generalization of Redfield's enumeration theorem [4, 5] which results in an expression for $g(n,m)$ as a function of $Z(S_m)$ and $Z(S_2)Z(S_{n-2})$. In order to present such a formula for $g(n,m)$, we require some special functions which depend on Redfield's cup operation.

As above let R be the ring of rational polynomials in the variables $a_1, a_2, a_3, \ldots$. For each positive integer r we shall define a function $J_r \colon R \to R$. We begin by defining $J_r(a_k^j)$ and it is convenient to do this by first considering a sequence of functions $d_1, d_2, d_3, \ldots$ which depend on r and k. For each $s = 1, 2, 3, \ldots$ we let

(35)
$$d_s = \begin{cases} a_{ks}/k & \text{if } s \mid r \text{ and } (r/s, k) = 1 \\ 0 & \text{otherwise} \end{cases}$$

Then $J_r(a_k^j)$ is given by

(36) $$J_r(a_k^j) = j! k^j Z(S_j; d_1, d_2, \ldots, d_j) .$$

For monomials we define J_r by

(37) $$J_r(\prod_{k=1}^{n} a_k^{j_k}) = \prod_{k=1}^{n} J_r(a_k^{j_k}) .$$

Next J_r is extended linearly to R. In particular

(38) $$J_r(Z(A)) = \frac{1}{|A|} \sum_{\alpha \in A} J_r(\prod_{k=1}^{n} a_k^{j_k(\alpha)}) .$$

Now we construct a ring from the collection $\mathcal{F}$ of all functions from R to R. For F_1 and F_2 in and p in R, addition is defined as usual by

(39) $$(F_1 + F_2)(p) = F_1(p) + F_2(p) .$$

For multiplication, we use the cup operation:

(40) $$(F_1 \cdot F_2)(p) = F_1(p) \cup F_2(p) .$$

It is easily seen that $(\mathcal{F}, +, \cdot)$ is a commutative ring. Furthermore $Z(S_m; J_1, J_2, \ldots, J_m)$ is an element of and when it is applied to the polynomial $Z(S_2) Z(S_{n-2})$ in R, the image is a polynomial in R whose coefficient sum is $g(n,m)$. This conclusion is summarized in the next theorem.

Theorem 4. The number of multigraphs $g(n,m)$ is the sum of the coefficients of the polynomial which is the image of $Z(S_2) Z(S_{n-2})$ under the function $Z(S_m; J_1, J_2, \ldots, J_m)$; symbolically

(41) $$g(n,m) = [Z(S_m; J_1, J_2, \ldots, J_m)(Z(S_2) Z(S_{n-2}))]_{a_i = 1} .$$

Some of the details in finding $g(5,4)$ using formula (41) are now sketched. First of all, we have

(42) $$Z(S_2) Z(S_3) = \frac{1}{12}(a_1^5 + 4a_1^3 a_2 + 2a_1^2 a_3 + 3a_1 a_2^2 + 2a_2 a_3)$$

and

(43) $$Z(S_4; J_1, J_2, J_3, J_4) = \frac{1}{24}(J_1^4 + 6J_1^2 J_2 + 8J_1 J_3 + 3J_2^2 + 6J_4) .$$

From formulas (35) through (38) we have the following results:

$$J_2(Z(S_2) Z(S_3)) = \frac{1}{12}(a_1^5 + 10a_1^3 a_2 + 15a_1 a_2^2 + 2a_1^2 a_3 + 2a_2 a_3 + 6a_1 a_4)$$

$$J_3(Z(S_2) Z(S_3)) = \frac{1}{12}(a_1^5 + 20a_1^2 a_3 + 4a_1^3 a_2 + 8a_2 a_3 + 3a_1 a_2^2)$$

$$J_4(Z(S_2)Z(S_3)) = \frac{1}{12}(a_1^5 + 10a_1^3a_2 + 15a_1a_2^2 + 30a_1a_4 + 2a_1^2a_3 + 2a_2a_3) \ .$$

From the definitions (32) and (33) of sums and products in we find

$$[J_1^4(Z(S_2)Z(S_3))]_{a_i=1} = 107$$

$$[6J_1^2J_2(Z(S_2)Z(S_3))]_{a_i=1} = 162$$

$$[8J_1J_3(Z(S_2)Z(S_3))]_{a_i=1} = 40$$

$$[3J_2^2(Z(S_2)Z(S_3))]_{a_i=1} = 69$$

$$[6J_4(Z(S_2)Z(S_3))]_{a_i=1} = 30 \ .$$

Therefore $g(5,4) = \frac{1}{24}(107+162+40+69+30) = 17.$

The great advantage of this method is, of course, that the only cycle index formulas needed are those of the symmetric groups. The computation of the functions $J_r(a_k^j)$ may be somewhat difficult, but these functions have many uses other than the one indicated here (see [5]).

REFERENCES

1. M. Hall, <u>The theory of groups</u>, New York, 1959.

2. F. Harary, <u>A seminar on graph theory</u>, New York, 1967.

3. F. Harary and E. Palmer, The enumeration methods of Redfield, <u>Amer. J. Math.</u>, 89 (1967), 373-384.

4. E. Palmer and R.W. Robinson, The matrix group of two permutation groups, <u>Bull. Amer. Math. Soc.</u>, 73 (1967), 204-207.

5. E. Palmer and R.W. Robinson, A generalization of Redfield's theorem. (to appear)

6. G. Pólya, Kombinatorische Anzahlbestimmungen für Gruppen, Graphen und chemische Verbindungen, <u>Acta Math</u>, 68 (1937), 145-254.

7. J.H. Redfield, The theory of group-reduced distributions, <u>Amer. J. Math.</u>, 49 (1927), 433-455.

CHARACTERIZATIONS OF 2-DIMENSIONAL TREES

Raymond E. Pippert and Lowell W. Beineke
Purdue University at Fort Wayne

The graphs known as trees are basic in graph theory and in applications to such fields as chemistry, electric networks, and game theory. There are many equivalent ways of defining trees, the most common being this: A tree is a graph which is connected and has no cycles. Figure 1 shows the trees with up to six vertices. Some equivalent definitions of a tree are the following: (i) A tree is a graph which is connected and has one more vertex than edge, and (ii) A tree is a graph which has no cycles and has one more vertex than edge. For these and some other characterizations see Berge [2] and Harary [3]. A less common definition or characterization is inductive: The graph consisting of a single vertex is a tree, and a tree with $n + 1$ vertices is obtained from a tree with n vertices by adding a new vertex adjacent to exactly one of the others.

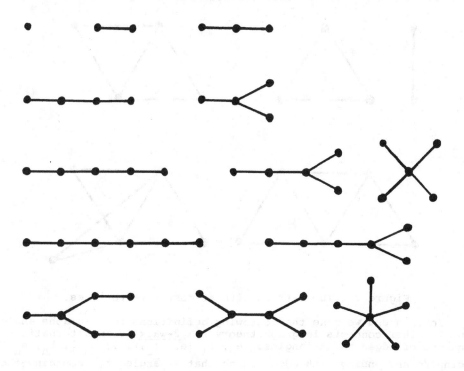

Figure 1. The trees with at most six vertices.

This inductive definition suggests an extension of the concept of tree, and for this we find the following definitions useful. (See Harary and Palmer [4] for similar definitions.)

A 2-dimensional graph or 2-graph consists of a nonempty set V and a collection C of nonempty subsets of V such that
i)   For each v in V, the set {v} is a member of C.
ii) Each set in C has at most three elements.
iii) Every nonempty subset of a member of C is in C.
In other words, a 2-graph is a simplicial complex of dimension at most 2. The three-element sets in C will be called triangles, the two-element sets edges, and the members of V vertices. (In this paper we may also use the term vertices for the one-element sets without confusion.)

A 2-dimensional tree or 2-tree is a 2-graph described as follows. The 2-graph consisting of an edge joining two vertices is a 2-tree, and a 2-tree with n + 1 vertices is obtained from a 2-tree with n vertices by adding a new vertex and the triangle which contains that vertex and two already adjacent vertices.

Figure 2 shows the 2-trees with up to five vertices; there are five with six vertices. Beineke and Pippert [1] have enumerated the labeled 2-trees, Harary and Palmer [4] the unlabeled. The object of this note is to provide further characterizations of 2-trees.

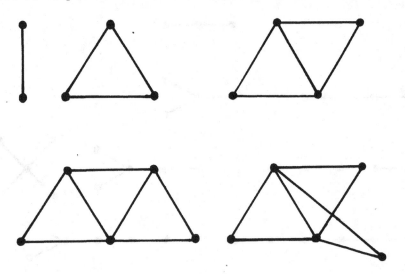

Figure 2. The 2-trees with at most five vertices.

To this end we make the following definitions for 2-graphs analogous to other concepts in graph theory. A 2-walk is an alternating sequence of edges and triangles, $e_0, t_1, e_1, t_2, \ldots, e_{n-1}, t_n, e_n$, beginning and ending with edges, such that triangle t_i contains the distinct edges e_{i-1} and e_i. This 2-walk is a 2-cycle if n > 0,

$e_n = e_o$, and all other elements of the sequence are distinct. A 2-graph is called 2-linked (the term 2-connected would be preferred, but that has another meaning) if it has no isolated vertices and every pair of edges are joined by a 2-walk.

A 2-linked component or 2-component of a 2-graph G is a maximal 2-linked subgraph of G. Clearly, the edges and triangles of G are partitioned by the 2-components, whereas the vertices need not be. A vertex can lie in more than one 2-component, and isolated vertices lie in none. We also note that if a triangle (but not the corresponding edges or vertices) is removed from a 2-linked graph and the result is not 2-linked, then each 2-component shares at least two of its vertices with other 2-components.

Several properties of 2-trees are obtained in the following theorem.

Theorem 1. Let G be a 2-tree with p vertices. Then
(a) Every pair of edges of G are joined by a 2-walk.
(b) G has $2p - 3$ edges.
(c) G has $p - 2$ triangles.
(d) G has no 2-cycles.

Proof. This is a straightforward induction proof. Certainly all four conditions hold in the 2-tree with two vertices. Assume they hold in all 2-trees with n vertices. The construction of a 2-tree with $n + 1$ vertices from a 2-tree with n vertices yields a structure having one additional triangle, two additional edges, no 2-cycles, and in which every pair of edges are clearly joined by a 2-walk. Therefore all 2-trees satisfy the four conditions.

The remainder of this note consists in determining which sets of the conditions (a), (b), (c), and (d) of Theorem 1 serve to characterize 2-trees. Three lemmas will be useful in establishing the theoems.

Lemma 1. Let G be a 2-graph with q edges ($q \geq 1$) and r triangles. Then
(i) If G has no 2-cycles, $q \geq 2r + 1$.
(ii) If G is 2-linked, $q \leq 2r + 1$.

Proof. (i) Assume G has no 2-cycles. If $r = 0$ the result is immediate, so we may assume that G has at least one triangle. We first show that G has a triangle with at least two edges on no other triangle. Suppose not; that is, suppose every triangle in G has at least two edges in common with other triangles. Begin at some edge e_o which lies on a triangle t_1 and follow this with another edge e_1 of t_1 which lies on another triangle t_2. Take another edge e_2 of t_2 on another triangle t_3. Continuing this process, we obtain a 2-walk, $e_o, t_1, e_1, t_2, \ldots, e_{n-1}, t_n, e_n$, in which a triangle or edge must eventually repeat. Since no two consecutive edge terms or triangle terms are the same, there must be a 2-cycle in G, which proves the assertion. It is now clear that two edges of G may be removed without reducing the number of triangles by more than one. The graph so obtained still has no 2-cycles, so the procedure described above may be repeated, continuing until we obtain a graph

which contains only one triangle. In this graph it is clear that
$q \geq 2r + 1$, and the general result is obtained by induction.

(ii) Now assume G is 2-linked. Fix an edge e. To each of
the other p - 1 edges, associate the triangle containing it in some
2-walk of fewest terms joining it with e. Each triangle can be
associated in this way iith at most two edges, so that $2r \geq q - 1$,
which completes the proof.

Corollary. Let G be a 2-graph with q edges $(q \geq 1)$ and r triangles. If G has no 2-cycles and is 2-linked, then $q = 2r + 1$.

Lemma 2. If G is a 2-linked 2-graph with p vertices and q
edges, then $q \geq 2p - 3$.

Proof. The result is clearly true for 2-graphs having two or three
vertices. Assume it is true for 2-graphs having fewer than p vertices and let G be a 2-graph with p vertices that is 2-linked.
We can assume that the removal of any triangle from G results in a
2-graph which is not 2-linked, since otherwise edges and triangles can
be removed in succession until that is the case.

Now remove a triangle from G. The resulting 2-graph has at
least two and at most three 2-components. Each of these 2-components
must have fewer vertices than G since otherwise more edges and triangles could have been removed without destroying the property of being 2-linked. Let the number of vertices and edges in these 2-compo-
nents be p_i and q_i for i = 1, ..., k (k = 2 or 3). Since each
2-component has at least two vertices in common with the others, we
have $\sum_{i=1}^{k} p_i \geq p + k$. By the induction hypothesis, $q_i \geq 2p_i - 3$, for
i = 1, ..., k. Thus

$$q = \sum_{i=1}^{k} q_i$$

$$\geq \sum_{i=1}^{k} (2p_i - 3)$$

$$\geq 2p + 2k - 3k$$

$$\geq 2p - 3.$$

This completes the proof.

We define an end triangle to be one in which one vertex is adjacent to only the other two vertices of the triangle. We observe that
a trivial induction argument shows that every 2-tree with at least
four vertices has at least two end triangles. The next lemma gives
other conditions under which this is true.

Lemma 3. Let G be a 2-graph with at least four vertices, none isolated. If G satisfies the four conditions (a) - (d) of Theorem 1,
then it has at least two end triangles.

<u>Proof</u>. Let G be a 2-graph with p vertices, none isolated, such that conditions (a) - (d) are satisfied. Then G is 2-linked, so every vertex lies on a triangle. Furthermore, no triangle can have three edges each lying on only one triangle. But G has no 2-cycles, so there are at least p edges each of which lies on only one triangle. It follows that, since G has only p - 2 triangles, at least two triangles must have two edges each lying on only one triangle.

Remove such a triangle and the corresponding two edges. If the resulting 2-graph G' were 2-linked, Lemma 2 would be contradicted, so G' is not 2-linked. Let v be the vertex common to the two removed edges and let e be the third edge of that triangle. Now G' cannot have an edge at v for otherwise there must have been a shortest walk in G joining it to e, and this walk would have to be in G' which would thus be 2-linked. It follows that v is isolated in G' so the removed triangle was an end triangle in G. Since there are two such triangles in G, the lemma is established.

It is now readily established that the four properties given in Theorem 1 serve to characterize 2-trees.

<u>Theorem 2</u>. Let G be a 2-graph with at least two vertices, none isolated. If G satisfies conditions (a) - (d) of Theorem 1, then it is a 2-tree.

<u>Proof</u>. The proof is inductive. The theorem is certainly true when there are 2 or 3 vertices. Assume it is true when there are p vertices, p ≥ 3. Let G be a graph with p + 1 vertices satisfying the hypotheses. By Lemma 3, it has an end triangle. Remove its vertex of degree 2 (and of course the triangle and edges incident with the vertex). What remains still satisfies the hypotheses and by the induction assumption is a 2-tree. It follows from the definition that G itself was a 2-tree, which suffices to complete the proof.

In fact, not all four of the properties are required for characterizing 2-trees, as the following theorem shows.

<u>Theorem 3</u>. Let G be a 2-graph without isolated vertices, and let p be the number of vertices, q the number of edges, and r the number of triangles. Then the following are equivalent.
   (1) G is a 2-tree.
   (2) Every pair of edges of G are joined by a 2-walk and $q = 2p - 3$.
   (3) Every pair of edges of G are joined by a 2-walk and $r = p - 2$.
   (4) G has no 2-cycles, $q = 2p - 3$, and $r = p - 2$.

<u>Proof</u>. That (1) implies each of (2), (3), and (4) is contained in Theorem 1. We proceed to show that each of (2), (3), and (4) implies (1).

First assume G satisfies (2). If p = 2 or 3, G is clearly a 2-tree. Suppose there is a 2-graph that satisfies (2) but is not a 2-tree. Let p be the minimum number of vertices in such a 2-graph, and let G be one of these which has the minimum number of triangles. Remove one triangle. The resulting 2-graph G' either has p - 2 triangles or is not 2-linked. Now if there are p - 2 triangles, then G' is a 2-tree. But in a 2-tree there cannot be three mutually adjacent vertices without the corresponding triangle. Hence, we may assume that G' is not 2-linked. Then since there are no isolated vertices, G' has either two or three 2-components; denote this number by k. Clearly for $i = 1, \ldots, k$, $q_i \geq 2p_i - 3$ and $\sum_{i=1}^{k} q_i = q$. Also $\sum_{i=1}^{k} p_i \geq p + k$, so that

$$2p - 3 = q$$

$$= \sum_{i=1}^{k} q_i$$

$$\geq \sum_{i=1}^{k} (2p_i - 3)$$

$$= 2 \left(\sum_{i=1}^{k} p_i \right) - 3k$$

$$\geq 2p - k.$$

From the inequality it follows that $k \geq 3$ and thus k must be 3. It also follows that $q_i = 2p_i - 3$ for $i = 1, 2, 3,$ and since no 2-component has more than two vertices in common with others, $\sum_{i=1}^{3} p_i = p + 3$. Now $r_i = p_i - 2$ by hypothesis, so that

$$r = \sum_{i=1}^{3} r_i + 1$$

$$= \sum_{i=1}^{3} (p_i - 2) + 1$$

$$= p - 2.$$

It has been established that in G every pair of edges are joined by a 2-walk and there are $2p - 3$ edges and $p - 2$ triangles. If G has no 2-cycles, then it is a 2-tree by Theorem 2. Suppose G has a 2-cycle. In this 2-cycle there are equal numbers of edges and triangles, and each of these triangles has an edge not in the 2-cycle. To each of the other edges associate the incident triangle in a 2-walk having the fewest terms joining it to an edge of the 2-cycle. Each triangle not in the 2-cycle is associated with at most two such edges, so it follows that $q \leq 2r$, which contradicts the facts that $q = 2p - 3$ and $r = p - 2$. Therefore G is a 2-tree.

Now assume that G satisfies (3). By Lemma 1, $q \leq 2r + 1$ so that $q \leq 2p - 3$. But Lemma 2 gives $q \geq 2p - 3$ and hence $q = 2p - 3$. That G is a 2-tree now follows from the preceding result.

Finally, assume that G satisfies (4). Let k denote the number of 2-components of G. Each 2-component is 2-linked and has no 2-cycles, so by the corollary to Lemma 1, the number of edges in each is one more than twice the number of triangles. Therefore, $q = 2r + k$. But since $q = 2p - 3$ and $r = p - 2$, k must be 1. This completes the proof.

Observing that property (a) of Theorem 1 is similar to that of being 2-linked, we restate Theorem 3 in the form:

<u>Theorem 3</u> (Alternate statement). Let G be a 2-graph with p vertices, q edges, and r triangles. Then the following are equivalent.
   (1) G is a 2-tree.
   (2) G is 2-linked and $q = 2p - 3$.
   (3) G is 2-linked and $r = p - 2$.
   (4) G has no isolated vertices and no 2-cycles, $q = 2p - 3$, and $r = p - 2$.

The reader can probably state other sets of conditions which are equivalent to the property of being a 2-tree, but most such would simply consist of different terminology. Harary and Palmer [4], however, give a more topological than combinatorial definition of a 2-tree, in that they require that G be 2-linked, have no 2-cycles, and be simply connected. They also show that a 2-graph is a 2-tree if and only if it is 2-linked, has no 2-cycles, and $p - q + r = 1$.

We conclude with several examples illustrating that no other set of the four properties characterizes 2-trees unless that set contains one of the sets listed in Theorem 3. The structures in Figure 3 are 2-graphs, without isolated vertices, which are not 2-trees although G_1 is 2-linked and has no 2-cycles, G_2 has no 2-cycles and $2p - 3$ edges, G_3 has no 2-cycles and $p - 2$ triangles, and G_4 has $2p - 3$ edges and $p - 2$ triangles.

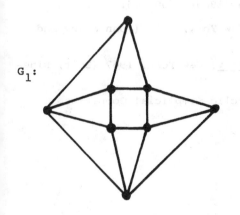

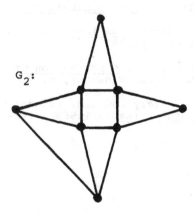

G_3: 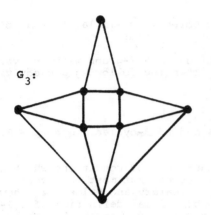 G_4: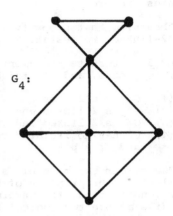

Figure 3

REFERENCES

1. L.W. Beineke and R.E. Pippert, The Number of Labeled k-Dimensional Trees, J. Combinatorial Theory, 6 (1969), 200-205.

2. C. Berge, The Theory of Graphs, New York, 1962 (John Wiley and Sons), p. 152.

3. F. Harary, A Seminar on Graph Theory, New York, 1967 (Holt, Rinehart, and Winston), p. 9.

4. F. Harary and E.M. Palmer, On Acyclic Simplicial Complexes, Mathematika 15 (1968), 115-122.

A COMBINATORIAL IDENTITY

C. Ramanujacharyulu, Bell Telephone Laboratories

Let n be a positive integer and

$$(n_1, n_2, \ldots, n_k) = \pi_k; \quad 1 \leq k \leq n$$

be a partition of n into k positive integers. Two partitions, π_k^1 and π_k^2, are said to be equivalent if one is a permutation of the other. Let S_k denote the set of all unequivalent partitions π_k of n, $k = 1, 2, \ldots, n$. Let π_k contain ℓ distinct integers among n_1, $n_2, \ldots, n_k$ and let r_i be the number of times the ith one of these distinct integers appears in π_k, $i = 1, 2, \ldots \ldots, \ell$.

Then we have the following identity.

Theorem.

$$\sum_{k=1}^{n} \sum_{\pi_k \in S_k} \frac{n!}{n_1! n_2! \ldots \ldots n_k!} \cdot \frac{1}{r_1! r_2! \ldots \ldots r_\ell!} \, n_1^{n_1-1} n_2^{n_2-1} \ldots \ldots n_k^{n_k-1} = (n+1)^{n-1}.$$

Proof. Following is a graph theoretic proof. The righthand side of the identity represents the number of trees [1] (i.e., connected graph without cycles) on $(n+1)$ vertices.

On the other hand the trees are counted by first finding the number of trees in which a fixed vertex, say x, appears with degree k (i.e., k edges or branches and only k are incident to it) and summing these numbers for $k = 1, 2, \ldots, n$.

Hence, all that is to be shown is that for a given partition,

$$\pi_k = (n_1, \ldots, n_k),$$

the term under double summation represents a component of the total number of trees where the degree of the vertex x is k. For this purpose, let the remaining n vertices be partitioned into k non-empty subsets $D_1, D_2, \ldots, D_k$ containing $n_1, n_2, \ldots, n_k$ vertices. The number of such partitions of n vertices is exactly

$$\frac{n!}{n_1! n_2! \ldots \ldots n_k!} \cdot \frac{1}{r_1! r_2! \ldots \ldots r_\ell!}$$

where $r_1, r_2, \ldots \ldots, r_\ell$ are as explained earlier. Consider an arbitrary tree on each of the subsets $D_1, \ldots, D_k$. A tree in D_i can be obtained in $n_i^{n_i-2}$ ways and a selection of trees on $D_1, \ldots, D_k$ subsets can be made in

$$n_1^{n_1-2} n_2^{n_2-2} \cdots \cdots \cdots n_k^{n_k-2}$$

ways. The vertex x is to be joined to one vertex and only one from each of the k trees so that the degree of x is k which can be done in exactly

$$n_1 n_2 \cdots n_k$$

ways.

Thus the number of trees where the vertex x has degree k, given a partition π_k of n is

$$\frac{n!}{n_1! n_2! \cdots n_k!} \cdot \frac{1}{r_1! r_2! \cdots \cdots r_\ell!} n_1^{n_1-1} n_2^{n_2-1} \cdots \cdots n_k^{n_k-1} \ .$$

Thus summing this over all unequivalent partitions π_k of S_k and over $k = 1, 2, \ldots, n$ we have all the trees on $(n+1)$ vertices which are all distinct. Hence the identity.

As J. Riordan ([2], p. 118) has communicated, this identity can also be established as follows:

Let $R_k = k^{k-1}$ (which is the number of rooted trees on k vertices); $T_n = n^{n-2}$ (the number of trees on n vertices) and Y_n the Bell polynomial. Then

$$Y_n(R_1, R_2, \ldots \ldots, R_n) = \sum_{k=1}^{n} Y_{n,k}(R_1, \ldots \ldots, R_n) \ ,$$

and

$$Y_{n,k}(R_1, R_2, \ldots \ldots, R_n) = \binom{n}{k} n^{n-1-k} k;$$

furthermore ([2], p. 96) we find:

$$b(b+n)^{n-1} = a_n = \sum_{k=0}^{n} \binom{n}{k} n^{n-1-k} k b^k$$

from which, putting $b = 1$, we get

$$(n+1)^{n-1} = \sum_{k=0}^{n} \binom{n}{k} n^{n-1-k} k = \sum_{k=1}^{n} Y_{n,k}(R_1, R_2, \ldots \ldots, R_n)$$

where $Y_{n,k}(R_1, R_2, \ldots \ldots, R_n)$ is the second summation in the identity.

Denoting the left-hand side of the identity by $T(n+1)$, the identity can be rewritten as follows:

$$T(n+1) = \sum_{k=1}^{n} \sum_{\pi_k \in S_k} \frac{n!}{n_1! n_2! \cdots \cdots n_k!} \cdot \frac{n_1 n_2 \cdots \cdots n_k}{r_1! r_2! \cdots \cdots r_\ell!} T(n_1) \ldots T(n_k)$$

which, when treated as a functional equation in integers, has the solution $T(n) = n^{n-2}$.

REFERENCES

1. O. Ore, <u>Theory of Graphs</u>, A.M.S. Colloq. Publ. Vol. 38, 1967.

2. J. Riordan, <u>Combinatorial Identities</u>, Wiley, New York, 1968.

AN APPLICATION OF GRAPH THEORY
TO SOCIAL PSYCHOLOGY

James E. Riley, Western Michigan University

Introduction. An important and interesting part of social psychology
is that of group dynamics. This area is concerned with the structure
of the relationships between individuals within a group and the changes such structures undergo. A graph may be employed in a very natural way to represent a group of people and some relationship which
may exist between certain pairs of individuals, whereby, individuals
are represented by vertices and the existence of an edge joining two
vertices would represent some relationship between the individuals.

In most cases the mere presence or absence of a relation is not
adequate to represent a given social structure. One may be concerned
with a group where a relation between two individuals may be categor-
ized as positive, negative, or nonexistent. Such a situation can be
treated with the aid of signed graphs, i.e., graphs whose edges are
designated positive or negative. It is precisely this case which is
investigated here with regard to three special problems, namely, balance, clustering, and credibility.

The Problem of Balance. Consider a group of individuals in which,
with regard to some issue, every two people agree, disagree, or their
relative attitudes on the subject are not known. Heider [3] considered this situation with a group of three people where he assumed
that for each pair of people there is either agreement or disagreement. Four possibilities exist as shown in Figure 1, where agreement
and disagreement are respectively exhibited by solid and broken lines.

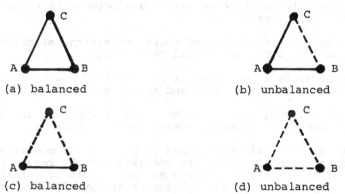

(a) balanced   (b) unbalanced

(c) balanced   (d) unbalanced

Figure 1

In (a), there is total agreement among the three people; Heider
called this system "balanced". He called the system (b) "unbalanced"
since the person C finds himself in the awkward position of agree-

ing with A and disagreeing with B while A and B are in agreement. In (c), persons B and C disagree but A has sided in with
B and opposes C. Heider felt there was consistency in this arrangement and again termed (c) as "balanced". He classified the system
(d), in which only disagreement existed, as unbalanced.

Cartwright and Harary [1] have generalized Heiders concept of
balance so as to include any system which can be represented by a
signed graph. The definition of balance given below is not á la
Cartwright-Harary, but it is equivalent, and preferable for the purposes of this paper.

A signed graph S is said to be <u>balanced</u> if its vertex set can
be partitioned into two subsets (one of which may be empty) so that
any edge joining two vertices within the same subset is positive,
while any edge joining two vertices in different subsets is negative.
The signed graph of Figure 2 is therefore balanced.

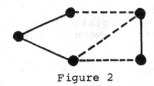

Figure 2

A social system which can be represented by a signed graph S
is said to be <u>balanced</u> if S is balanced. Some social psychologists
believe that within any social system of the type under consideration
there is a "tendency toward balance," implying that in an unbalanced
system there is stress or tension and a tendency for the system to
readjust (such as certain individuals within the group changing their
point of view) so as to relieve this tension. Thus there is a tendency for the group to split into two factions such that within a fac-
tion there are only positive relations and between factions there are
only negative relations.

The following definition and characterizations of balanced signed graphs are due to Cartwright and Harary [1].

A path or cycle in a signed graph is called <u>positive</u> if it has
an even number of negative edges and <u>negative</u> otherwise.

<u>Theorem 1</u>. A signed graph S is balanced if and only if all paths
joining the same pair of vertices have the same sign.

<u>Proof</u>. Let S be a balanced signed graph. If S contains only
positive edges, then all paths are positive and the result follows
immediately. Otherwise, the vertex set of S can be partitioned into two nonempty sets V_1 and V_2 such that any edge joining two
vertices of V_i, i = 1,2, is positive while every other edge is
negative. It is now easily observed that any path joining two vertices in V_i, i = 1,2, contains an even number of negative edges and
so is positive while any path joining a vertex of V_1 with one in
V_2 has an odd number of negative edges and so is negative.

Conversely, assume that S is a signed graph having the property that all paths joining the same pair of vertices have the same sign. Without loss of generality, we take S to be connected for otherwise we can treat each connected component of S individually. Let v be a vertex of S, and define the set V_1 to consist of v and all vertices u of S such that there is a positive path between u and v. Let V_2 denote all other vertices of S. There can be no positive edge of the type (v_1, v_2), $v_1 \in V_1$, $v_2 \in V_2$, for this leads to the existence of a positive path between v and v_2, contradicting the fact that $v_2 \notin V_1$. Also there can be no negative edge of the type (u,w), $u,w \in V_i$, $i = 1,2$, for all paths from v to u have the same sign as those paths between v and w. If P is a path between v and u not containing (u,w), then $P,(u,w),w$ is a path between v and w having the opposite sign of P. If every path between v and u contains (u,w) then let P' be one such path. A path between v and w can then be produced by deleting (u,w) and u from P'. In either case, a contradiction arises. Thus, $V_1 \cup V_2$ is an appropriate partition so that S is balanced.

Theorem 2. A signed graph S is balanced if and only if every cycle of S is positive.

Proof. Let S be a balanced signed graph and suppose S has a negative cycle C. The cycle C therefore contains an odd number of negative edges. Let u and v be any two distinct vertices of C. The cycle C induces two edge-disjoint paths between u and v, one necessarily containing an even number of negative edges and the other containing an odd number of negative edges. This implies that there is a negative path joining u and v as well as a positive path, and this contradicts Theorem 1. Hence, every cycle of S is positive.

Assume now that S is a signed graph in which every cycle is positive. If S were not balanced, then, by Theorem 1, there would exist two vertices u and v and two paths P' and P" joining them, one of which is positive and the other negative. It is not difficult to see that P' and P" together induce a negative cycle which contradicts the hypothesis. Thus, S is balanced.

As an illustration of the preceding, consider a social system in which we have a group of people with friendliness and unfriendliness occurring between certain pairs of individuals; furthermore, assume the existence of a rumor which has two basic forms, one true the other false. Suppose also that one would pass on the rumor to a friend in the same form as he had received it but would change the form if he were to pass on the rumor to someone to whom he was unfriendly. Theorem 1 states, then, that if the system is balanced, each person will hear only one version of the rumor regardless of the manner by which it reached him, and by Theorem 2, any person who starts a rumor will have it returned to him in the same form as he originally knew it.

The Problem of Clustering. It is the belief of some social scientists that it may be unnatural to expect a tendency toward balance in a social system in which both positive and negative relations exist. A related but alternative theory is to anticipate a "clustering" of the people into several groups (not necessarily two) where positive relationships occur only within a group and negative relationships occur only between different groups. Once again this leads to a natural application of signed graphs.

A signed graph S is said to be clusterable if its vertex set V can be partitioned into subsets, called clusters, so that every positive edge joins vertices within the same subset and every negative edge joins vertices in different subsets. The following result is due to J.A. Davis [2].

Theorem 3. A signed graph S is clusterable if and only if S contains no cycle with exactly one negative edge.

Proof. Assume S is a clusterable signed graph, and let C be a cycle of S. If C contains only vertices from a single cluster, then all edges of C are positive. If C contains vertices from two or more clusters of S, then C contains at least two edges joining different clusters, i.e., at least two negative edges. However, in either case C does not contain exactly one negative edge.

Conversely, let S be a signed graph containing no cycle with exactly one negative edge. Define a relation on the vertex set V of S such that two vertices u and v are related if either u = v or u and v are joined by an all-positive path, i.e., a path all of whose edges are positive. This relation is readily seen to be an equivalence relation on V and, as such, induces a partition of V into equivalence classes, where two distinct vertices belong to the same equivalence class if and only if these vertices are joined by an all-positive path. We must now show that these classes are clusters so that S is clusterable. There can be no positive edge of the type (u,v), where u and v are in different equivalence classes, for any two vertices joined by an all-positive path belong to the same class. Likewise, there can be no negative edge of the type (u,v), where u and v belong to the same equivalence class, for if u and v belong to the same class, an all-positive path exists between them. This path together with the negative edge (u,v) produces a cycle with exactly one negative edge, and this is a contradiction. Therefore, S is clusterable.

The Problem of Credibility. Thus far it has been assumed that every relation occurring within a given social system is symmetric. However, if person A has a positive relationship toward person B, there is no reason to believe that B has a positive relationship toward A. Such a structure can be represented in a natural way by directed graphs.

Recall that for balanced signed graphs, all paths joining the same pair of vertices have the same sign. We now consider the corresponding situation for directed signed graphs. Before proceeding further, however, a few definitions are needed.

A semi-cycle is an alternating sequence of vertices and edges $a_1, x_1, a_2, x_2, \ldots, a_n$, where the vertices are distinct except $a_1 = a_n$, x_i is either the directed edge (a_i, a_{i+1}) or the direct-

ed edge (a_{i+1}, a_i), and there is exactly one vertex a_i incident
from two edges in the sequence. Thus every semi-cycle can be express-
ed in the form a_1, (a_1, a_2), a_2, ..., (a_{m-1}, a_m), a_m, (a_{m+1}, a_m), a_{m+1},
..., (a_1, a_{n-1}), a_1. (The directed graph of Figure 3 is a semi-cycle).

The <u>converse</u> D' of a directed graph D is the directed graph
having the same vertex set as D and such that (a,b) is an edge of
D' if and only if (b,a) is an edge of D.

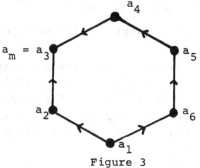

Figure 3

As expected, a path, cycle, or semi-cycle in a directed graph is
called positive or negative depending on whether it contains an even
or odd number of negative edges. A directed signed graph D is
called <u>credible</u> if for every two vertices u and v of D, all
paths from u to v have the same sign. Thus a credible directed
signed graph represents a social system (as it did in an earlier example) in which if a rumor is started by person A and is received
in one of two possible forms by person B, then B will always hear
the rumor in the same form; thus, the rumor has "credibility". A
characterization of credible directed signed graphs is now presented.

<u>Theorem 4</u>. A directed signed graph D is credible if and only if
every semi-cycle of D is positive.

<u>Proof</u>. Let D be a credible directed signed graph, and let
C: a_1, (a_1, a_2), a_2, ..., (a_{m-1}, a_m), a_m, (a_{m+1}, a_m), a_{m+1}, ...,
(a_1, a_{n-1}), a_1 be a semi-cycle of D. One sees that P_1: a_1,
(a_1, a_2), a_2, ..., (a_{m-1}, a_m), a_m and P_2: a_1, (a_1, a_{n-1}), a_{n-1}, ...,
(a_{m+1}, a_m), a_m are two paths from a_1 to a_m and so are either both
positive or both negative, i.e., each of P_1 and P_2 contains an
even number of negative edges or an odd number of negative edges. In
either case, C contains an even number of negative edges and is
therefore positive.

Conversely, suppose every semi-cycle of a directed signed graph
D is positive. Let P and Q be two paths in D from vertex u
to vertex v, where u and v are arbitrary distinct vertices of
D. Let u_1, u_2, ..., u_k be the vertices common to P and Q,
written in the order of their occurrence, where, then, $u_1 = u$ and
$u_k = v$. For each i, $1 \le i \le k-1$, either (u_i, u_{i+1}) is an edge

common to P and Q, or u_i and u_{i+1} lie on a semi-cycle C_i of D, and the number of negative edges of C_i on P and on Q are both even or both odd. Thus the number of negative edges on P from u_i to u_{i+1} and the number of negative edges on Q from u_i to u_{i+1} for $1 \le i \le k-1$ are either both even or both odd. Hence, P and Q have the same sign.

We conclude with the following result.

Theorem 5. A directed signed graph D is credible if and only if its converse D' is credible.

Proof. This is a direct consequence of the observation that C': a_1, (a_1, a_2), a_2, $\ldots$, (a_{m-1}, a_m), a_m, (a_{m+1}, a_m), a_{m-1}, $\ldots$, (a_1, a_{n-1}), a_1 is a semi-cycle of D' if and only if C: a_1, (a_2, a_1), a_2, $\ldots$, (a_m, a_{m-1}), a_m, (a_m, a_{m+1}), a_{m+1}, $\ldots$, (a_{n-1}, a_1), a_1 is a semi-cycle of D, and C and C' are both positive or both negative.

REFERENCES

1. D. Cartwright and F. Harary, Structural Balance: a generalization of Heider's Theory, Psychological Review, 63 (1956), 277-293.

2. J.A. Davis, Clustering and structural balance in graphs, Human Relations, 20 (1967), 181-187.

3. F. Heider, Attitudes and Cognitive organization, Journal of Psychology, 21 (1946), 107-112.

A TOPOLOGICAL INFLUENCE:
HOMEOMORPHICALLY IRREDUCIBLE GRAPHS

M. James Stewart, Lansing Community College

A collection of p points (0-simplexes), and q arcs (1-simplexes), joining certain pairs of points (not necessarily distinct) which is imbedded in 3-space in such a way that every intersection of arcs occurs only at some of the p points is a finite geometric simplicial 1-complex, or simply, a 1-complex. (Some examples are shown in Figure 1). Two 1-complexes $\mathfrak{C}_1$ and $\mathfrak{C}_2$ are homeomorphic if there exists a one-one bicontinuous mapping from $\mathfrak{C}_1$ onto $\mathfrak{C}_2$. Geometrically, this means that each of $\mathfrak{C}_1$ and $\mathfrak{C}_2$ can be continuously deformed into the other (see Figure 1). Also, the relation "is homeomorphic with" is an equivalence relation, and therefore partitions all 1-complexes into equivalence classes.

(a)   (b)   (c)

Figure 1. Geometric 1-complexes: (b) is
homeomorphic with (c) but not with (a).

Now it is well known that every graph can be realized as a 1-complex, and that every 1-complex can be embedded in 3-space. So for every graph there is at least one associated geometric 1-complex. In the light of this observation, it is convenient to make the following definition: two graphs are said to be homeomorphic if their associated geometric 1-complexes are homeomorphic. Thus from this definition it follows that the partition of all geometric 1-complexes into equivalence classes under the "is homeomorphic with" relation induces a partition on the set of all graphs into equivalence classes. Within each such equivalence class of graphs, we designate a graph H as <u>homeomorphically irreducible</u> if out of all members of this class, H has a minimal number of vertices. Before describing some of the properties of homeomorphically irreducible graphs, the following definitions are helpful.

A <u>walk of length n</u> joining two vertices v_o and v_n of a graph is an alternating sequence of $n+1$ vertices and n edges v_o, v_ov_1, v_1, v_1v_2, v_2, $\ldots$, v_n; if all vertices are distinct, the walk

is a path. Here v_1, v_2, ..., v_{n-1} are called <u>interior vertices</u> of
the walk. A vertex is called <u>suppressible</u> if it has degree 2 and
the vertices to which it is adjacent are not themselves joined by an
edge.

Now let us consider to what extent two homeomorphic graphs G_1
and G_2 can differ. Since their associated geometric 1-complexes
are topologically equivalent, both G_1 and G_2 must have the same
number of vertices of degree different from two (for these correspond
to the locally non-Euclidean points of the associated 1-complexes)
but may differ in the number of vertices of degree two. More specifically, every vertex in G_1 or G_2 of degree two which is suppressible corresponds to an interior point of an arc in the associated 1-
complex. Since this arc could just as well correspond simply to a
single edge used in place of the suppressible vertex and its two adjacent edges, we see that homeomorphic graphs may differ only in the
number of suppressible vertices they possess. Hence two homeomorphic
graphs G_1 and G_2 are in fact identical if they possess the same
number of suppressible vertices.

If H is a homeomorphically irreducible graph, by definition it
has the smallest number of vertices out of all graphs in its equivalence class, and so H must possess no suppressible vertex. Hence
any other homeomorphically irreducible graph G in this same class
must be identical to H (since they have the same number of suppressible vertices). Thus we have shown

<u>Theorem 1</u>. A graph is homeomorphically irreducible if it has no sup-
pressible vertices; furthermore, under homeomorphisms, every such
graph is the unique representative of its equivalence class.

<u>Corollary</u>. A graph is homeomorphically irreducible if it has no ver-
tices of degree two.

There are several ways of characterizing homeomorphically irreducible graphs. We conclude with the following result.

<u>Theorem 2</u>. For a nonempty graph G, the following are equivalent:
    (a) G is homeomorphically irreducible.
    (b) Every vertex of degree two in G lies on a triangle.
    (c) If A denotes the adjacency matrix of G, then the matrix
    $A^2 + A^3$ has no diagonal entry equal to 2.

GRAPH THEORY AND "INSTANT INSANITY"

Joan Van Deventer
Albion College and Michigan State University

In the past few years, various games using multicolored cubes have become popular; a particular example is the puzzle called "Instant Insanity", which lends itself to a solution by means of graph theory. In this and similar puzzles, four unit cubes have faces colored arbitrarily with four colors, such that each color appears on at least one face of each cube. The challenge is to stack the cubes in a vertical 1 x 1 x 4 rectangular prism so that, if possible, each of the four colors appears on each long side of the prism. There exist colorings which yield no solution, while other colorings yield one or more distinct solutions, that is, up to a permutation of the order of cubes. A trial-and-error method is, in most cases, unsatisfactory since it is conceivable that one could try all of the 41,472 possibilities, but not arrive at a solution until the very last try.

Problems concerning the stacking of colored cubes have been discussed by Busacker and Saaty [2], who exhibited a solution to a particular cube coloring problem using methods of graph theory, and by Brown [1], who gave a solution to the "Instant Insanity" puzzle by assigning positive integers to the four colors and solving an associated problem in number theory. The present paper extends the graph theory methods used in [2], and also presents a detailed solution to the "Instant Insanity" problem.

To analyze the means of solving such puzzles, let us assume that a solution exists. Consider the completed rectangular prism standing with the long sides vertical. (Fig. 1) Here the letters represent the four colors blue, green, white, and red. The front and back long sides will be considered as one set, as will the left and right long sides. Since this arrangement is assumed to provide a solution, the set of front and back faces consists of two square faces of each color. If the colors are represented by vertices, and the relationship of "opposite" on a given cube is represented by an edge connecting the vertices representing the opposite colors, the graphical representation is a graph with four vertices and four edges, with multiple edges and loops possibly appearing. Furthermore, each vertex has degree two (a loop at a vertex giving that vertex degree two). In addition, the edges may be labeled according to which cube they represent. For convenience, the cubes are numbered 1, 2, 3, and 4 from the base of the prism. The left and right sides may be portrayed in the same way. Consequently any solution may be represented by two graphs of degree two, each having four vertices, and four edges, numbered one to four, possibly with loops or multiple lines. (Fig. 2)

B	G	W	R	4
G	W	R	W	3
W	B	B	G	2
R	R	G	B	1

b l f r

Figure 1

left-right   front-back

Figure 2

We next exhibit a solution for the puzzle "Instant Insanity."
Consider the four cubes before positioning. Following the above method, it is possible to represent each cube as a graph with four vertices and three edges, allowing multiple edges and loops. (Fig. 3)

Cube 1:   Cube 2:   Cube 3:   Cube 4:

Figure 3

Now consider the graph formed when the graphs of each of the
four cubes are superimposed on the same set of four vertices. (Fig.
4) If these four cubes yield a solution, it should be possible to
find two line-disjoint subgraphs of degree two and order four, one
representing front-back color-oppositeness, and one representing
left-right color-oppositeness, each containing edges numbered one
through four. Two such subgraphs are shown in Figure 5, and can be
shown to be unique.

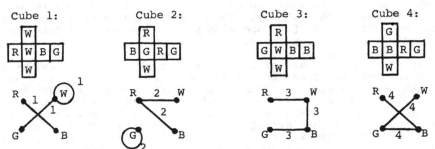

Figure 4

Figure 5

These two constitute the graph for the solution, which may be realized by placing cube one in such a way that the opposite sides which do not appear in either of the subgraphs become the "buried" sides. The opposites which appear on the front and back sides determine the placement of all the other cubes. Letting f, b, l, r stand for front, back, left, and right, label the subgraphs, beginning with the cube 1 edge, in the manner shown in figures 6 and 6A. Stacking the cubes in the indicated way gives the solution shown in figure 7.

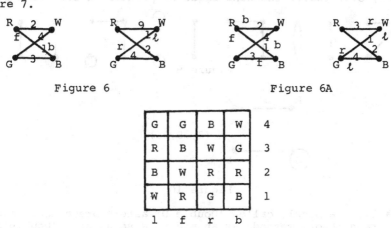

Figure 6   Figure 6A

G	G	B	W	4
R	B	W	G	3
B	W	R	R	2
W	R	G	B	1
	1	f	r	b

Figure 7

We now return to the general problem of coloring cubes with four colors. Note that by removing labels from the possible subgraphs of degree two, there are five non-isomorphic possibilities. (Fig. 8)

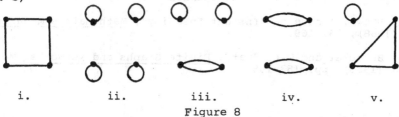

i.   ii.   iii.   iv.   v.

Figure 8

One possible extension of the problem would be to consider n cubes in n colors. Clearly if n > 6, the restriction that each color appear on each cube would have to be lifted.

A second question might be to consider the basic colorings of a cube where each of the four colors appears at least once on each cube. There are six non-isomorphic forms from which a cube might be colored, (Fig. 9). Brown [1] assigns the numerical values 1, 2, 3, 5 to red, white, blue, and green respectively, and represents "oppositeness" by the product of those numerical values related to the opposite colors. Consequently each cube is expressed in terms of a 1 x 3 row matrix. He then forms a 4 x 3 matrix using those associated with the four cubes, and searches for two disjoint sets of four numbers, one from each row, whose product is 900. Since a solution depends on two of each color on the front and back sides of the prism, as well as left and right, the numerical values for the front and back long sides must be $1^2 \cdot 2^2 \cdot 3^2 \cdot 5^2 = 900$. Using Brown's method to represent each graph of a cube (Fig. 10) it is possible to consider which colorings do give solutions. Since there are over three hundred thousand combinations, practicality requires a computer. At this point, over twelve thousand pages of printed output are expected.

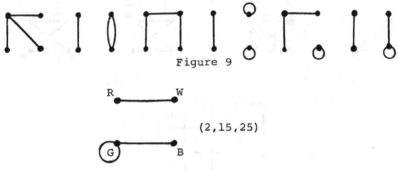

Figure 9

R———————W

(2,15,25)

G———•

B

Figure 10

Finally, the puzzle called "Double Disaster" presents a chance to apply graph theory methods as an aid to a solution. "Double Disaster" consists of eight unit cubes, colored as in "Instant Insanity", where the object is to form a 2 x 2 x 2 cube where each of the four colors appears on each face, including the top and bottom.

REFERENCES

1. T.A. Brown, A note on "Instant Insanity", Mathematics Magazine, 41 (1968), 167-169.

2. R.G. Busacker and T.L. Saaty, Finite Graphs and Networks, McGraw-Hill (1965), pp. 153-155.

ARC DIGRAPHS AND TRAVERSABILITY

Curtiss E. Wall, Olivet College

Introduction. The line-graph L(G) of a graph G is a graph whose point set can be placed in one-to-one correspondence with the line set of G in such a manner that adjacency is preserved. Line-graphs and their properties (particularly those involving traversability) have been studied by Chartrand [1,2]. In [3], Harary and Norman introduced the analogous concept of the arc digraph (or line digraph). The arc digraph L(D) of a digraph D is one whose point set can be put in one-to-one correspondence φ with the arc set of D so that point u is adjacent to (from) point v in L(D) if and only if arc φ(u) is adjacent to (from) arc φ(v) in D. In Figure 1, a digraph D and its arc digraph L(D) are shown.

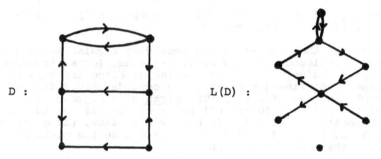

D :   L(D) :

Figure 1

Regular arc digraphs. For a point u of a digraph D we denote by id u and od u the indegree and outdegree of u. We define D to be inregular of degree r if id u = r for all points u of D; outregular digraphs are defined analogously. A regular digraph of degree r is one which is both inregular and outregular of degree r. In addition, we define D to be arc-regular of degree r if for each arc x = uv of D, id u = od v = r. In this section, we discuss regularity and arc-regularity as it is involved with arc digraphs. We begin by stating a result due to Harary and Norman [3].

Theorem. Let D be a (weakly) connected digraph. Then D ≅ L(D) if and only if D is inregular of degree 1 or outregular of degree 1.

The following result is a direct consequence of the definition.

Proposition 1. The arc digraph L(D) of a connected digraph D is regular if and only if D is arc-regular.

It is obvious that if a digraph is regular, then it is also arc-regular of the same degree. The converse of this statement (for connected digraphs) is also true, as we now show. Let D be a connected, arc-regular digraph of degree r > 0, and let u be an arbitrary point of D. We consider two cases:

Case 1. od u > 0. Then there exists an arc x = uv. Since D is arc-regular, id u = od v = r. Thus, there is an arc y = wu so that id w = od u = r. Hence id u = od u = r, and D is regular of degree r.

Case 2. od u = 0. We must have id u > 0 since D is connected. Therefore, there exists an arc y = wu. Since D is arc-regular, id w = od u = r, but this is a contradiction, implying only Case 1 is possible.

This constitutes a proof of the following result.

Theorem 1. A connected digraph D is regular if and only if D is arc-regular.

The following two corollaries are now immediate.

Corollary 1a. If D is a connected digraph, L(D) is regular if and only if D is regular.

Corollary 1b. If D is a connected digraph, L(D) is arc-regular if and only if D is arc-regular.

We now see that the situation regarding regular arc-digraphs is not entirely analogous to regular line-graphs, for a line-graph L(G) is regular if and only if G is regular or biregular bipartite. Another property of line-graphs which is considerably different for arc-digraphs is that of girth. The girth of a digraph D is the length of the smallest (directed) cycle of D; if D has no cycles, the girth remains undefined. In nearly all cases, the girth of a line-graph L(G) is 3 and does not depend on the girth of G. Such is not the case for digraphs.

Theorem 2. The arc digraph L(D) of a digraph D has girth g if and only if D has girth g.

Proof. Let D be a digraph having girth g. Certainly, then, L(D) contains a cycle of length g. Suppose, however, that L(D) contains a cycle of length k < g. Let v_1, v_2, ..., v_k, v_1 be such a cycle in L(D). Then there exists a sequence x_1, x_2, ..., x_k, x_1 of arcs in D such that x_i is adjacent to x_{i+1}, i = 1,2,...,k-1, and x_k is adjacent to x_1. This arc sequence produces a cyclic sequence of k on fewer points of D, contradicting the fact that D has girth g.

Eulerian arc digraphs. A connected digraph D is said to be eulerian if there exists a closed (directed) trail containing all arcs of D. It is well known that a connected digraph is eulerian if and only if id v = od v for all points v of D. We further define D to be arc-eulerian if for every arc x = uv, id u = od v. As before, we have the following observation.

Proposition 2. The arc digraph L(D) of a connected digraph D is eulerian if and only if D is arc-eulerian.

For eulerian graphs G, the line-graphs L(G) are also euler-

ian. This situation, however, does not occur for digraphs; for example, the digraph D of Figure 2 is eulerian while L(D) is not.

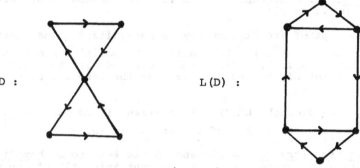

Figure 2

Although, as we have seen, regularity and arc-regularity are
equivalent for digraphs, the property of being eulerian neither implies nor is implied by the property of being arc-eulerian. In Figure 3, the digraph D_1 is arc-eulerian but not eulerian while D_2
is eulerian but not arc-eulerian where, furthermore, $D_2 = L(D_1)$.

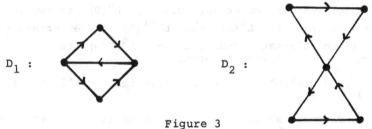

Figure 3

If a digraph D is both eulerian and arc-eulerian, then a further remark concerning D can be made.

<u>Theorem 3</u>. If a digraph D is eulerian and arc-eulerian, then D
is regular.

<u>Proof</u>. Choose an arbitrary point v_1, then there exists some trail
$v_1v_2, v_2v_3, \ldots, v_nv_1$ which contains all of the arcs of D. Since
D is eulerian,

$$id(v_1) = od(v_1), \ id(v_2) = od(v_2), \ \ldots, \ id(v_n) = od(v_n). \quad (1)$$

Also D is arc-eulerian, therefore

$$id(v_1) = od(v_2), \ id(v_2) = od(v_3), \ \ldots, \ id(v_n) = od(v_1). \quad (2)$$

Combining (1) and (2) we have

$$id(v_1) = od(v_1) = od(v_2) = id(v_2) = \ldots = id(v_n) = od(v_1) = id(v_1).$$

Therefore, D is regular of degree r.

The next result follows from the above theorem.

Corollary 3a. If a digraph D and its arc digraph L(D) are eulerian, then D is regular.

The concept of arc digraph may be generalized. The iterated arc
digraph is denoted $L^n(D)$. In addition, Proposition 2 may be generalized. For a trail P, id(P) and od(P) denote the indegree of
the initial point of P and outdegree of the terminal point of P,
respectively.

Theorem 4. The digraph $L^n(D)$ is eulerian if and only if $id(P_n) =$
$od(P_n)$ for every trail P_n of length n in D.

Hamiltonian arc digraphs. A digraph D is said to be hamiltonian if
there exists a (directed) cycle in D containing all of the vertices
of D. Clearly, if D is eulerian then L(D) is hamiltonian. The
converse of this statement is also true as was shown by Kastelyn, for
example, in [4]. This result and Theorem 2 imply the following result.

Proposition 3. If L(D) is eulerian and hamiltonian, then D is
regular.

If D is a regular digraph, then D is eulerian and, by Corollary 1a, L(D) is also eulerian. Clearly $L^n(D)$ is regular and
therefore eulerian. If $L^n(D)$ and $L^{n-1}(D)$ are eulerian, then
$L^{n-1}(D)$ is arc eulerian. Consequently, $L^{n-1}(D)$ is regular. These
remarks imply the following theorem.

Theorem 5. The digraphs $L^n(D)$ are eulerian for all n, if and
only if D is regular.

Since D eulerian implies L(D) is hamiltonian, we have the
following

Corollary 5a. The digraphs $L^n(D)$ are hamiltonian for all n if
and only if D is regular.

REFERENCES

1. G. Chartrand, Graphs and their associated line-graphs, Ph.D.
    Dissertation, Michigan State University (1964).

2. G. Chartrand, On hamiltonian line-graphs, Trans. Amer. Math. Soc.
    134 (1968), 559-566.

3. F. Harary and R.Z. Norman, Some properties of line digraphs,
    Rend. Circ. Mat. Palermo 9 (1960), 161-168.

4. P. Kastelyn, A soluble self-avoiding walk problem, Physica 29
    (1963), 1329-1337.

ecture Notes in Mathematics

isher erschienen/Already published

Bitte wenden / Continued

Beschaffenheit der Manuskripte
Die Manuskripte werden photomechanisch vervielfältigt; sie müssen daher in sauberer Schreibmaschinenschrift geschrieben sein. Handschriftliche Formeln bitte nur mit schwarzer Tusche eintragen. Notwendige Korrekturen sind bei dem bereits geschriebenen Text entweder durch Überkleben des alten Textes vorzunehmen oder aber müssen die zu korrigierenden Stellen mit weißem Korrekturlack abgedeckt werden. Falls das Manuskript oder Teile desselben neu geschrieben werden müssen, ist der Verlag bereit, dem Autor bei Erscheinen seines Bandes einen angemessenen Betrag zu zahlen. Die Autoren erhalten 75 Freiexemplare.

Zur Erreichung eines möglichst optimalen Reproduktionsergebnisses ist es erwünscht, daß bei der vorgesehenen Verkleinerung der Manuskripte der Text auf einer Seite in der Breite möglichst 18 cm und in der Höhe 26,5 cm nicht überschreitet. Entsprechende Satzspiegelvordrucke werden vom Verlag gern auf Anforderung zur Verfügung gestellt.

Manuskripte, in englischer, deutscher oder französischer Sprache abgefaßt, nimmt Prof. Dr. A. Dold, Mathematisches Institut der Universität Heidelberg, Tiergartenstraße oder Prof. Dr. B. Eckmann, Eidgenössische Technische Hochschule, Zürich, entgegen.

Cette série a pour but de donner des informations rapides, de niveau élevé, sur des développements récents en mathématiques, aussi bien dans la recherche que dans l'enseignement supérieur. On prévoit de publier

1. des versions préliminaires de travaux originaux et de monographies

2. des cours spéciaux portant sur un domaine nouveau ou sur des aspects nouveaux de domaines classiques

3. des rapports de séminaires

4. des conférences faites à des congrès ou à des colloquiums

En outre il est prévu de publier dans cette série, si la demande le justifie, des rapports de séminaires et des cours multicopiés ailleurs mais déjà épuisés.

Dans l'intérêt d'une diffusion rapide, les contributions auront souvent un caractère provisoire; le cas échéant, les démonstrations ne seront données que dans les grandes lignes, Les travaux présentés pourront également paraître ailleurs. Une réserve suffisante d'exemplaires sera toujours disponible. En permettant aux personnes intéressées d'être informées plus rapidement, les éditeurs Springer espèrent, par cette série de »prépublications«, rendre d'appréciables services aux instituts de mathématiques. Les annonces dans les revues spécialisées, les inscriptions aux catalogues et les copyrights rendront plus facile aux bibliothèques la tâche de réunir une documentation complète.

Présentation des manuscrits
Les manuscrits, étant reproduits par procédé photomécanique, doivent être soigneusement dactylographiés. Il est recommandé d'écrire à l'encre de Chine noire les formules non dactylographiées. Les corrections nécessaires doivent être effectuées soit par collage du nouveau texte sur l'ancien soit en recouvrant les endroits à corriger par du verni correcteur blanc.

S'il s'avère nécessaire d'écrire de nouveau le manuscrit, soit complètement, soit en partie, la maison d'édition se déclare prête à verser à l'auteur, lors de la parution du volume, le montant des frais correspondants. Les auteurs recoivent 75 exemplaires gratuits.

Pour obtenir une reproduction optimale il est désirable que le texte dactylographié sur une page ne dépasse pas 26,5 cm en hauteur et 18 cm en largeur. Sur demande la maison d'édition met à la disposition des auteurs du papier spécialement préparé.

Les manuscrits en anglais, allemand ou français peuvent être adressés au Prof. Dr. A. Dold, Mathematisches Institut der Universität Heidelberg, Tiergartenstraße ou au Prof. Dr. B. Eckmann, Eidgenössische Technische Hochschule, Zürich.